COMPLETE BOOK OF
Cordwood Masonry Housebuilding

COMPLETE BOOK OF
Cordwood Masonry Housebuilding

▲

The Earthwood Method

ROB ROY

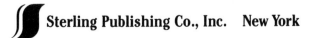

Sterling Publishing Co., Inc.   New York

Library of Congress Cataloging-in-Publication Data

Roy, Robert L.
    Complete book of cordwood masonry housebuilding : the earthwood method / Rob Roy.
       p.   cm.

    Includes bibliographical references and index.
ISBN 0-8069-8590-9
1. Log-end houses—Design and construction—Amateurs' manuals.
2. Earth sheltered houses—Design and construction—Amateurs' manuals.
TH4818.W6R67 1992
690'.837—dc20   91-41313

                                                                         CIP

    10   9   8   7   6   5   4   3   2   1

Published by Sterling Publishing Company, Inc.
387 Park Avenue South, New York, N.Y. 10016
©1992 by Robert L. Roy

Portions of this work are adapted from
Cordwood Masonry Houses, ©1981 by Robert L. Roy
and *Earthwood* ©1984 by Robert L. Roy
both published by Sterling Publishing Company
Distributed in Canada by Sterling Publishing
c/o Canadian Manda Group, P.O.Box 920, Station U
Toronto, Ontario, Canada M8Z 5P9
Distributed in Great Britain and Europe by Cassell PLC
Villiers House, 41/47 Strand, London WC2N 5JE, England
Distributed in Australia by Capricorn Link Ltd.
P.O. Box 665, Lane Cove, NSW 2066

Manufactured in the United States Of America

Sterling ISBN 0-8069-8590-9

For Rohan . . .
Son, friend, fellow skier and mortar stuffer.

"The time has come," the Royrob said,

"To talk of many things:

Of lime—and spuds—and cordwood logs—

Of why a chainsaw sings.

And what if log-ends shrink a lot?

And what's a 'Ship with Wings?' "

Acknowledgments:

Thanks to all the special people who helped contribute to the cordwood literature in general and this book in particular: Cordwood Jack Henstridge, George Barber, Bill Tishler, Richard Flatau, Harold Johnson, Thomas and Anne Wright, Geoff Huggins and Louisa Poulin, Jeff Corra, Aggie and Instant Bob Evans, Kim Hildreth, Paul Mikalauskas, Sam Davis, Deb and Ed Burke, I.R. and Dawn Lafon, Cliff Shockey, Sam Felts and Kevin Rencarge, Irene Boire, Betsy Pugh, Jerry Mitchell, Ken Campbell, Ron and Bea Farrell, Seigfried Blum and all the past students at Earthwood who have gone on to build their own homes. Thanks to Northeast Publications, Plattsburgh, NY, for permission to reprint the picture of Rohan's bicycle generator, Jim LaMar for some timely work in the darkroom, and Charlie Nurnberg and John Woodside for their patience. Most of all, Jaki, my loving thanks to you for making it all possible, including the late nights spent getting this sent off by deadline.

Contents

Color section follows page 128.

Introduction

cord'wood ma'son • ry, n., that building technique in which short logs (see *log-ends*) are laid widthwise in the wall within a special mortar matrix.

log-ends, n. pl., blocks, butts, or ends of logs used in the construction of a cordwood masonry wall; typically 6" to 24" in length.

The reader now has the vocabulary to discuss an increasingly popular method of low-cost home construction.

But watch out. These dry definitions hide a potentially habit-forming activity. I know. I've been hooked on cordwood masonry for 15 years, and it is a tough habit to kick, but why bother trying? True, it is mind-expanding and mildly intoxicating, but it is also invigorating and spiritually satisfying. Cordwood masonry has provided me and my family with low-cost energy-efficient shelter, introduced us to hundreds of wonderful people from all over the Western Hemisphere, and even helped to supplement our income.

Over those sixteen years, however, I have found people who are inclined to dismiss cordwood masonry as a fanciful or—even worse—*funky* way to build. Sometimes, the reaction is decidedly negative. Some experienced carpenters and masons have insisted: "You just can't build that way, son." Some affirm that the wood will rot; others that the mortar will loosen and reduce the wall to a heap of rubble. The concept of laying up wood with mortar seems to violate long-established tenets of construction.

Less skeptical types, especially those discussing the concept while standing inside our Earthwood house, will allow me to wax on for a while about the "Five E" advantages of cordwood masonry described in Chapter One, namely: Economics, Energy efficiency, Ease of construction, Esthetics, and Ecological harmony. Sometimes I have to avoid "overselling" the technique, lest the idea be rejected as "too good to be true."

Like any construction method, cordwood masonry has disadvantages as well as advantages. The drawbacks will not be downplayed in this book; but rotting of wood will not be found amongst them in a properly constructed cordwood wall. Still, readers sharing this fear may need some reassurance.

Rot occurs when wood is kept in a damp condition, allowing microorganisms to maintain a favorable climate for growth. These conditions are avoided in a properly built cordwood house by using sound wood to begin with, keeping the wood off the ground, and protecting the walls with a good overhang. Log-ends will get wet in a driving rain, but they breathe easily along their end grain, and soon dry out. The microorganisms never get a foothold. A split rail fence will endure a thousand storms with no sign of rot for the same basic reason: the rails breathe.

Also, cordwood walls last because each log-end is protected from being in contact with its neighbor by the mortar joint. The mortar is strong in lime which helps to preserve the wood. There are existing cordwood buildings in Europe about 1000 years old, and I have seen many in Canada over 100 years old where the log-ends would be good enough to use again in a new house.

It has been eight years since the publication of my book *Earthwood: Building Low-Cost Alternative Houses* (Sterling, 1984) and twelve years—Jimmy

Carter was president—since the appearance of *Cordwood Masonry Houses* (Sterling, 1980). Cordwood masonry was just beginning to take off in the early 80s, the result of those books and two or three others, several popular articles in magazines like *Harrowsmith* and *The Mother Earth News*, and local publicity and word of mouth about successful owner-built homes throughout North America.

It is safe to say that there are at least three times as many cordwood houses today as there were a decade ago. And it seems that every cordwood builder learns something new to add to the literature, so there is a whole lot more information to share. Most of this information, I'm happy to say, is in the form of new approaches and good ideas that worked: success stories. Other lessons learned have been in what not to do.

These lessons may be even more valuable to today's would-be homebuilder. There are so many potential errors lurking out there that we at least ought to avoid those that others have already made.

The upshot is that it is time to completely update the information on the subject. This book combines the information of the two books cited above, but the text of Part One is almost totally rewritten from *Cordwood Masonry Houses*, and Part Two has been thoroughly reviewed, corrected, and updated. A great deal of new information is included, especially regarding types of wood, mortar mixes, and how long to season the wood before building. Most of the case histories are new.

Part One, *Cordwood Masonry*, describes the cordwood masonry technique in detail, including discussions of all three major styles of construction, types of woods to use, mortar mixes, and special effects. There are several examples of successful owner-built homes from around the country. Part Two, *Earthwood*, takes the reader step by step through the building of the earth-sheltered cordwood home where I live and work with my wife and building partner, Jaki, and our sons, Rohan, 15, and Darin, 6. Although Earthwood is a round home, the basic lessons of construction from the foundation to the roof are appropriate to almost any shape.

Rob Roy
Earthwood Building School
West Chazy, New York

ONE
CORDWOOD
MASONRY

1

Overview of Cordwood Masonry

HISTORICAL PERSPECTIVE

In an article titled "Poor Man's Architecture" appearing in *Harrowsmith* No. 15, writer David Square says, "Curiously, the origin of the technique remains mysteriously obscure. In Siberia and in the northern areas of Greece, stackwall structures estimated to be 1,000 years old are still standing. Yet no one is certain where it all began." [1]

Among the more documented cases of early cordwood construction in the United States are two examples where, independently, New York State residents migrated west and built cordwood buildings in Walworth County in southern Wisconsin at about the same time. One was a house completed in 1849, according to Paul B. Jenkins, writing in *The Wisconsin Magazine of History* in December of 1923,[2] although further research by William H. Tishler suggests an 1857 date.[3] The house was built by David Williams, a farmer from western New York, and composed, according to Jenkins, of wood "cut, sawed, and split into sticks fourteen inches in length, exactly such sticks as are used for all kitchen cook fires." Jenkins estimated that 20,000 "sticks" (log-ends) were used and adds, "The work of preparing the wood must have been great and prolonged, and to this day (1923) Williams' descendants repeat the family tradition that every stick was sawed... with a common bucksaw." The house was torn down in 1950, despite attempts to save it by Jenkins. Fortunately, a specimen of the wall was removed and remains intact today at the Webster House Historical Museum in Elkhorn, Wisconsin.[4]

At Fontana, in the vicinity of the Williams house, was a cordwood masonry gristmill built about 1850 by Carlos Lavalette Douglass, who had come to Walworth County from Cataraugus County, New York. The building was torn down about 1913.[5]

At about the same time or perhaps a little earlier, cordwood masonry was already common in Canada's Ottawa Valley, and along the St. Lawrence in Quebec. Many old cordwood structures can be found in these areas. (They are not always easy to find, as people tended to plaster over or clapboard the old cordwood walls. In some cases, this may have been to cover up deterioration in poorly built walls, but in many cases people were trying to cover up evidence that they lived in "poor people's housing.")

In the late 19th and early 20th centuries, Wisconsin's Door County peninsula became the hotbed of cordwood construction in the States, with some 40-odd examples of houses and barns known. The best remaining example, however, lies in the hamlet of Jennings in Oneida County in north central Wisconsin. Built in 1899, the large structure was simultaneously a general store, saloon, bunkhouse, post office, and home of the John Mecikalski family. The building was added to the National Register of Historic Places in 1984, and was fully restored, thanks to a grant by the Kohler Foundation, between 1985 and 1987.

At the same time of the early Wisconsin cordwood construction—the mid-19th century—the

1-1. An old abandoned cordwood house in the town of Clay Banks in Door County, Wisconsin. Note the clapboards covering the side walls and rolled roofing on the gable end wall. *(Photo credit: William H. Tishler)*

1-2. Detail of an old "stovewood" barn near a Swedish settlement in Bayfield County, Wisconsin. *(Photo credit: William H. Tishler)*

1-3. (above) The Mecikalski store in Jennings, Wisconsin, before restoration…

1-4. (right) …and after. *(Both photos credit: William H. Tishler)*

technique was being used in Sweden. William H. Tishler says, "The validity of a Scandinavian origin was given further support in a 1979 (Tishler's) interview with an 80-year-old mason in northern Wisconsin. The builder of several cordwood structures, he confirmed learning this construction method from his father who acquired the technique in his native Sweden before emigrating to America."[6]

In short, the origin of cordwood masonry is lost in time, at least for now. My own view is that cordwood masonry is such a simple idea that it may have been spontaneously conceived by many different people at different times and at different places. I was once riding along a country road in search of a missing dog. I noticed a porch attached to the front of a farmhouse, the roof of which was supported by four sturdy posts. The people living there had stacked firewood between the house and the corner posts, and between the other posts, except for a clearway that had been maintained between the two central posts. I imagine that these woodpiles sheltered the front door from the winter winds very nicely—which may have been part of the intention—and provided close and easy access to the fuel supply.

Given the situation, it would have taken little more imagination to see the possibilities of these cordwood walls as shelter.

Jack Henstridge conceives of the invention of cordwood construction in much the same way. As he tells it,

The most precious thing that early man had was fire. He soon found out that unless he had a good supply of dry fuel the fire would go out, so he piled the wood up around a central fire. This worked great. Man not only had a good supply of fuel, but he soon found out that he could get in between the fire and the wood pile and keep warm. He then found out that if he built the pile high enough, he could lay long sticks over the top of it and some broad leaves over the sticks and the rain would not put the fire out. He also found that if he poked mud between the sticks in the pile, he could keep the wind out, too. All he had to do was to pile his circle of firewood so that the inside was even. He didn't have any saws or axes back then, so the outside was a veritable pincushion. His enemies could not get in at him. Man had invented the "Cordwood Wall" or more accurately the "Firewood Wall." He did not need the cave anymore.[7]

What I like about Jack's little story is its complete plausibility. In fact, cordwood masonry may have been discovered and rediscovered many times by events not unlike those described in Jack's fantasy. Jack further speculates that the Viking people brought the cordwood idea to North America about a thousand years ago. It is generally accepted that the Norsemen spent some time at L'Anse au Meadow on the northern tip of Newfoundland, where remains have been found that Jack likens to a round cordwood house, built with clay instead of mortar. The remains are too far gone to make this determination with certainty, but in discussing the matter Jack and I began to consider the following hypothesis: What if the Vikings built a round house with cordwood and plenty of clay and then, before roofing, built a huge bonfire within the structure—or around its edges—and "vitrified" the walls, virtually welding wood and clay into one long-lasting monolithic wall. The ends of the cordwood would burn, but the fire would go out with the drawing away of the heat by the wet clay. A similar technique was used to vitrify earthen forts in ancient Britain.

Perhaps the most tantalizing story I've heard about early cordwood masonry construction is that instructions for building a cordwood wall were found on a 3000-year-old clay tablet at Knossos in Crete. Unfortunately, I've lost the reference. If any reader can throw any light on this one, I'd be grateful if you'd get in touch.

Over its long history, cordwood masonry has been known by many names: Stovewood architecture; Firewood walls; Wood masonry infilling; Stackwall construction; Stackwood. There are probably others, but don't panic. They all refer to the same building technique where short logs are stacked up in a wall like firewood. To avoid confusion, I'll stay with *cordwood* as the general term for this type of construction, and *log-ends* as the name for the blocks, butts, logs, ends, or pieces that are laid up in a cordwood wall. The reader should watch out for alternative terms in other books and articles, however.

THE FIVE E'S OF CORDWOOD MASONRY

The second little pig was on the right track. His house of sticks would have withstood any blowhard wolf had he paid attention to basic principles of cordwood masonry construction. And his house would have enjoyed other advantages not obtained by his elder brother, the conservative Practical Pig, advantages which can be conveniently thought of in terms of five E's:

(1) *Economics* Cordwood houses are of high quality and low cost. Little maintenance reduces the ongoing cost, too. We were able to build Earthwood in 1981-82 for under $10 per sq ft for materials. Our summer cottage, Mushwood, featured as a case history in Chapter Six, is just nearing completion as I write, and is also coming in at a materials cost of about $10 per sq ft: 900 sq ft for about $9000. Some of our resourceful former students at the building school have done even better than this, and some have spent a lot more, usually when they have contracted a lot of the work out.

(2) *Energy Efficiency* A properly constructed cordwood house with walls the right thickness for the climate will be easy to heat in winter and stay naturally cool in summer. The key is the double-mortar joint. At no point in the wall is the inner mortar joint in direct contact with the outer mortar joint, so there is no "energy nosebleed." An insulated cavity exists between the mortar joints, adding greatly to the R-value. The wall, then, combines excellent insulation with phenomenal thermal mass.

(3) *Ease of Construction* The cordwood technique is easy to learn and easy to do. Aggie Evans—a young grandmother—laid 90 percent of the log ends at their 1400 sq ft round house near us, while her 14-year-old son, Jeff, mixed the mortar, which is really the most strenuous job. The individual log-ends weigh only a few pounds each.

(4) *Esthetics* A cordwood wall combines the warmth of wood with the pleasing relief and interest of stone masonry. The very few individuals I've met who have said that they would not like to have a cordwood wall exposed in their home usually take an equally dim view of any rustic building style, including logs, stonework, and exposed beams. They like smooth surfaces. Well, everybody's different.

(5) *Ecology* Cordwood masonry is kind to the environment. "Waste" woods are often perfectly acceptable to use when cut into log ends. When the house finally gives itself up to the unalterable march of time, it does so gracefully. Most of the materials can return peacefully to the surroundings from which they came.

GENERAL REMARKS

Although the possibilities for individual initiative and imagination when building with cordwood seem to be unlimited, construction generally falls within one of three distinct categories, each of which will be discussed separately and at length. The purpose of this section is to give a general description of cordwood construction, and to compare some of the advantages, disadvantages and special considerations of the three distinct approaches to cordwood housing. This will give the reader a wider perspective when reading the chapters that are devoted to each individual style.

Before launching this discussion, two short but important notes might be appropriate.

(1) I don't intend to favor any single method over the others. Rather than pursue a meaningless discussion over which method is "best"—there is no correct answer to this—I encourage people to choose the one that most appeals to them and best suits their individual circumstances…and then to go out and build!

(2) It should be kept in mind that the three distinct construction techniques can be combined in various ways—there's no need to stick exclusively to one method. A house should reflect the owner's character and no two people are the same. I have seen hundreds of different cordwood homes—in person, in books and magazines, and through correspondence—and no two are alike, except in their use of log-ends in the walls. Cordwood is like modelling clay: it just begs to be molded and caressed into a thousand different shapes. It is tempting to lecture about the economic and energy savings of cordwood, tell people how long these houses last, how beautiful they are, how easy to build, but the one concept which is hardest to impart to a prospective builder is that cordwood masonry is fun and good for the soul. Don't take my word for it.

The three different ways in which cordwood masonry can be employed in building are: (1) within a *post-and-beam framework*, (2) within a log or log-end framework of *built-up corners*, and (3) as a *load-supporting curved wall*.

The unique cross section of a cordwood wall is common to all three styles. The log-ends themselves establish the width of the wall, as the ends of the logs are exposed on both the interior and exterior.

The mortar, however, does not pass directly through the wall. There is an unseen cavity, filled with insulation, in the middle of the wall. As an example, the mortar matrix of a 16" thick wall is typically composed of two 5" mortar joints separated by 6" of loose fill insulation.

POST-AND-BEAM FRAMEWORK

By this method, the framework, not the log-ends, does the load supporting. Our first house, Log End Cottage, is a good example of this style (*see* Chapter 3). The main advantage of this method is that the entire frame can be built, and the roof put on, before any cordwood masonry is done. Log-ends can be drying while the frame is going up. Once the roof is on, cordwood work can proceed, even in inclement weather. With the other styles, it is necessary to cover the work each night to guard against rain entering the insulated cavity.

The down side is that the thickness of the cordwood wall is limited to the dimensions of the posts and beams in the framework. We used 9" log-ends at Log End Cottage, for example, because the old hand-hewn barn timbers we obtained had dimensions varying from 8" by 8" to about 9" by 10". In the south, this would probably be perfectly acceptable, but, as we have found out, northern heating requirements dictate cordwood walls with the equivalent of 16" of white cedar. Jeff Corra, of Parkersburg, West Virginia, overcame this problem by building a "double-wide" framework, as discussed in Chapter 3.

The post-and-beam style is a must in earthquake zones. Without this important skeleton, a masonry wall, which is strong on compression but weak on tension, can oscillate and may come tumbling down in a strong quake.

CURVED WALLS

Cordwood homes can be almost any curved shape: oblong, oval, spiral, round, freestyle. But the easiest—especially when it comes to roofing—and the most efficient in terms of labor and materials is round.

A round house of any given perimeter will

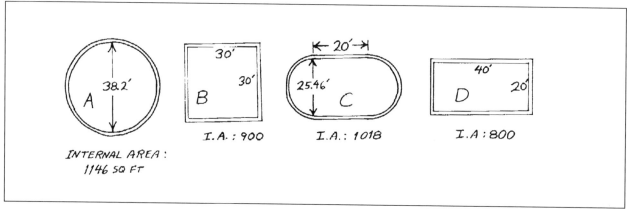

1-5. The perimeters of all four of these house shapes are the same: 120 ft. The curved-wall shapes are a much more efficient use of cordwood.

enclose 27.3 percent more area than the most efficient of the rectilinear shapes, the square. Compared with the more commonly shaped house that Western peoples build, where the length is about twice the breadth, the gain of building space is about 43 percent. Think of it: 43 percent more space for the same amount of labor and materials. The other engineering species of this planet—such as birds, bees, and beavers—know this instinctively, with no need for a course in geometry.

Another advantage is that a round house will be easier to heat than a square or rectangular house with the same floor area. The economy of wall area mentioned above is part of the reason, of course, but the lesser wind resistance offered by a curved wall is also important. And if a radiant heat source, such as a woodstove, is placed at the center of the structure, there will be no cold corners, as all points on the circumference will be equidistant from the heat source.

The primary disadvantage of any of the curved wall shapes is that the project is more at the mercy of the weather than the post and beam house which can get its roof on early. Walls have to be covered each night, unless there is a zero percent chance of rain. Another possible disadvantage is the lack of acceptance of a round house by others besides the builder. This could be the builder's banker, the builder's spouse, or even the building inspector, where there is strict zoning.

Can log-ends be load-supporting? After all, there is very little bond between wood and mortar. The answer is a resounding "You betcha!" On compression, the walls are very strong. Even in a worst case

scenario, where the wood and mortar both shrink—which scenario should not occur if you keep reading carefully—the resulting wall would be a perfectly constructed "dry stone" wall, with close fitting and wonderfully interlocking log-ends and pieces of mortar.

There are dry-stone cottages in the Western Isles of Scotland that have been occupied continuously for over 400 years and they have not benefited from a drop of mortar in their construction, no bond at all between the stones except that universal bonding agent: gravity.

The compression strength of a round cordwood house is phenomenal. Wood and mortar are both strong on compression, and the wall supports itself all around the circle. It can't fall inward because of the *keystone principle*. Means of stopping the wall from falling outwards are discussed in Chapter 5.

But the final proof is in the pudding. "Earthwood" is now over ten years old. The two-story cylindrical structure supports an earth roof which, when fully saturated and then subjected to a heavy spring snow, weighs over 200,000 lbs. About half of this load—over 50 *tons*—is carried in compression by the cordwood walls themselves. The bottom courses of cordwood masonry also carry the considerable load of two stories of a 16" thick wall of wood and mortar. There is no detectible strain on either the walls or any of the other structural components of the building. And, at 150 lbs per sq ft, our maximum roof load is three times the combined dead load and snow load of 50 lbs per sq ft demanded by our local building code (Plattsburgh, NY, near Montreal).

1-6. An example of the built-up corners style of cordwood masonry.

BUILT-UP CORNERS (STACKWALL)

The built-up corners method, also known as "stackwall," is an old style of construction and has something in common with each of the other methods.

The cordwood is load-supporting (although many builders tie the corners together with top plates), but the technique is also that of masonry infilling, this time between laid-up corners of longer log-ends or six-by-six timbers, instead of between posts and beams. The advantage of the stackwall system is that a rectilinear structure can be built completely out of cordwood without the need for a post-and-beam framework. This is faster, easier, and usually cheaper than the post-and-beam method, though probably not quite as strong per inch of wall thickness. This slight disadvantage is more than made up for by the use of thick walls. The thick walls serve as a better insulation and thermal mass for those contemplating a cordwood structure in a northern climate. In Canada, 24-inch thick walls are not uncommon.

2
Construction

This chapter deals with considerations applicable to all three styles of cordwood masonry, including foundation options, the mortar mix, wood selection and drying, barking and splitting the wood, and, finally, building the cordwood wall. Considerations unique to each style will be covered in each style's respective chapter.

FOUNDATION OPTIONS

The foundation for a cordwood masonry house should be a rigid masonry foundation. This can take several forms. My own personal preference (and, incidentally, Frank Lloyd Wright's favorite foundation system) is the floating slab. By this method, a slab of concrete "floats" on a thick "pad" of good percolating material such as coarse sand, gravel, or crushed stone. This method is described in Chapter Five, Curved Walls, with a step-by-step case history of how we actually built such a foundation in Part Two, Earthwood.

A poured wall down below frost level is okay, either in combination with a basement space or not, but there are a couple of drawbacks. Firstly, it is the most expensive method, as it really should be done professionally. The one case where I know of owner-builders doing their own poured foundation for an underground home ended up with a "concrete blow-out" or failure of the forming materials. Five cubic yards of concrete had to be cleaned up, and fast. Spoiled their whole day. Secondly, the cordwood wall should not be thicker than the foundation upon which it sits, because of potential problems with wall instability. Most poured walls are only 8" thick, and an 8" cordwood wall just doesn't make it in northern climes. Might be okay in Georgia.

Laying up concrete blocks, particularly by the surface-bonding method, is a much more viable alternative to the poured wall. It is cheaper by far, do-able by the inexperienced owner-builder if basic care is taken, and the walls can be up to 16" thick, as you will see in Part Two.

Another good foundation method for those long on time and stones, but short on money, is a field-stone foundation wall, taken down to frost level. The reader is advised to consult the library or book stores to find one of the many available how-to reference works dealing specifically with stone masonry.

Another low-cost foundation method has been highly touted in the past for cordwood masonry construction, but does not seem to have passed the test of time. I refer to the ditch and gravel berm foundation method discussed in *Cordwood Masonry Houses*, and recommended by the Northern Housing Committee of the University of Manitoba in their book *Stackwall: How to Build It* (U. of Manitoba, 1977.) Railroad ties or pressure-treated timbers are laid on a berm of crushed stone, and the cordwood masonry proceeds immediately upwards from there. I have heard from Canadian students passing through our Earthwood Building School that some of these foundations have suffered from differential settling, causing major cracks in the cordwood wall after a while. This might be avoided if the indigenous soil has good percolation and the railroad ties are tamped down extremely well into the crushed stone

22

berm, but my advice is to use a more rigid frost-proof foundation method.

Similarly, I do not endorse a pillar foundation, with heavy beams or railroad ties joining the pillars. This is still not rigid enough. Vibration or deflection of the timbers will cause the cordwood wall to crack.

AVOID BASEMENTS

This is a good place to interject my diatribe on "basements." Avoid them. They are dark, damp, expensive, low-grade (no pun intended) space. We know. We built one at Log End Cottage. It was over one third the cost of the house and provided about 5 percent of the usage.

Basement space is very often dark, damp, dingy, and of little practical value. Its most useful purpose is to store garden produce. Its more common use is to store junk. Most functions, *including heating*, are best enclosed in the house proper, not in a basement. In a nutshell, basements are not cost-effective, require familiarity with an additional structural system (which is why most owner-builders contract their basement out, usually at high cost relative to the house) and provide low-quality space for almost every activity short of mushroom propagation.

If extra space is needed, go with a floating slab and a two-storey house. This is cheaper and will yield a better-quality space. Another option, which we exercised at Earthwood, is to spend the extra money on waterproofing, insulation, windows and doors, and transform the "basement" into a high-quality "earth-sheltered" space, with plenty of warmth, light and natural ventilation. How to upgrade the quality of below-grade space is discussed in detail in Part Two.

SAWDUST MORTAR

In cordwood masonry, the mortar matrix is equal in importance to the log-ends themselves, both structurally and in terms of appearance. Over a 16-year period, we have used several different mortar mixes, refining and improving over that time.

Our first efforts at Log End Cottage were not particularly successful. The mortar was hard and strong,

but shrinkage cracks developed between practically every pair of adjacent log-ends. Luckily, the construction style was post-and-beam and the heavy timber frame took all of the load, so the cottage stands happily today. The worst problem is by the front entrance, where years of pounding by a three-inch-thick, solid-wood door has loosened up the mortar in the adjacent cordwood panel.

Having worked for a mason in Scotland, I knew that cementitious materials—such as concrete, plaster, and mortar—shrink when they dry too fast, and crack when they shrink. The problem is exacerbated in a cordwood wall. The dry log-ends are trying to draw moisture out of the mortar. We tried to arrest the moisture loss by hanging wet towels over the work, but this met with little success. We may have slowed moisture loss to the air, but those darned log-ends were taking the Law of Entropy far too seriously and the mortar still dried out too fast. We altered our mix with only a little success. We were down to the last of about 20 cordwood "panels" in the house. The other 19 carried on shrinking and cracking.

"Genius comes from desperation," it is said, or should be. Or, in the words of the *auld* Scottish proverb: "When you're lucky, muck'll do for brains." Take your pick. In any case, we'd heard that Cornell University had been experimenting with sawdust-based concrete, and we figured we had nothing to lose by fooling around with the concept ourselves. I don't know what Cornell's goal was—it sure wasn't cordwood masonry—but we wondered if the sawdust would decrease shrinkage in some way, perhaps in the way that "air-entrained" concrete works.

The experiment worked, and (I like to fantasize) changed the course of cordwood masonry forever. But now, after years of reflection and further experimentation, I believe that the success of the sawdust mortar comes more from its ability to retard the set than from the creation of little cells which stop cracks in their tracks. My perception is that the sawdust acts as millions of little sponges, each storing a significant amount of moisture. As the mortar tries to dry, moisture is given back into the cement from these little saturated sawdust reservoirs. The mortar—or "mud" in the mason's vernacular—can still be scratched with the fingernail after three days and is not fully hard until five days or more after laying up.

Some masons, whose experience has been with stone, bricks or blocks, may cringe at the introduction of organic matter into a mortar mix. Good masons take care to keep topsoil, grass, and leaves out of their mud. But the brick mason can soak his hod of bricks prior to laying them up in a wall, so that there is no problem with dry bricks robbing moisture from the mud. Cordwood masons cannot safely use the same method, although some have tried. If dry log-ends are soaked in water, they will begin to swell. Then, there are two possibilities. Firstly, they will continue to swell in the wall as the mortar tries to cure. They will actually break up the wall, as will be discussed later. The second possibility is that they will swell and, if the timing is right, stop swelling before they are actually laid up. In this scenario, the wall will not break up, but, later on, the wood will shrink back to its original size, leaving exaggerated shrinkage gaps around each log-end.

On hot dry days with extreme drying conditions, it may be useful to give the log-ends a *quick dip*—in and out of a water vessel, no more—to prevent rapid loss of mortar moisture. Do not allow the wood to soak up water, which could cause the kind of problems discussed in the last paragraph.

Does the sawdust weaken the wall? No. By retarding the set, it actually strengthens the wall by eliminating shrinkage cracks. What happens to the sawdust in mortar after the wall is fully set? It dries out and shrinks a little bit, leaving little air gaps in the wall filled with dry shrivelled up sawdust. Is this a problem? No, in fact, such a mortar is bound to be better as insulation because of the millions of little sawdust-filled gaps.

The sawdust to be used in the mortar should be from a sawmill where softwoods are being cut. You can also use the sawdust created during the cutting of your log-ends, unless they are dense hardwoods. The trouble with hardwood sawdust, such as the post oak sawdust we used during a workshop in Texas, is that the particles are small, hard, and dense. They do not absorb and store moisture the way that, say, northern white cedar or pine sawdust does. Also, the small hard particles of oak seemed to make the mortar grainier and harder to work with, less "plastic," so to speak. Do not use the fine powdery sawdust that comes from a cabinetmaker's shop.

Woods vary around the country, of course, and the terms "hardwood" and "softwood" can be misleading. Poplar sawdust—though poplar is technically a "hardwood"—may be acceptable, while some dense "softwoods" may be hard and grainy. The important thing is to have a sense of the qualities that you are looking for—and avoiding—in your sawdust.

The sawdust should be passed through a half-inch (½") screen and straight into a soaking vessel. The screen eliminates undesirable shavings and bark in your mix. I use an open-topped 55-gallon drum for my soaking container. An old bathtub would also work well. The sawdust should be thoroughly soaked in water—overnight at least—before being introduced into the mix. Do not use unsoaked sawdust, as the dry sawdust will actually work against the intended purpose. It will rob the moisture from the mix instead of acting as a reservoir for replenishing the moisture.

Harold Johnson, a Canadian living near us in New York, owned a pulp mill in Canada. Rather than use sawdust as an admixture in his mud, he used wood pulp which he turned into a slurry in a 55-gallon drum. He used a large electric mixer to disintegrate the pulpwood blocks. The experiment was a success; his mortar did not shrink. The slurried pulp, it seems, retarded the set as well as the soaked sawdust. The Johnson home is featured in Chapter Six.

In September of 1991, Rohan and I visited two structures built by Roland Daoust of Herdman, Quebec. Roland used a very impressive mortar made of two parts of a specially refined vermiculite called "mortar aggregate" and one part masonry cement. The mortar was hard, strong and extremely lightweight. It seems to take pointing very well. Roland does not use an insulated cavity with his walls, but I think that the R-value would be greatly increased with the insulation—the same vermiculite would be excellent—and much mortar mixing would be saved. Roland feels that a three to one ratio would also work. An interesting feature of the mortar is that you can actually cut it with a chain saw. There is no sand.

SAND

Given a choice, I prefer a "sugar" (finer-grained) sand to a coarse grainy sand. In some areas, this is known as mason's sand. Bricklayers use it. I have found that a coarse sand is harder to make "plastic." It tends to crumble while handling it, instead of staying together in a nice flexible ball. When the finer sugar sand has been unavailable, I have made do with the coarser stuff, but have found it necessary to add a full

2-1. This recording studio in Athelstan, Quebec was made with a mortar of two parts vermiculite and one part masonry cement. The sound qualities of the studio are said to be excellent.

extra shovel of lime in with each mix to get a good workable mud of plastic consistency.

To give fair play to a totally different point of view, I must mention that cordwood builder and writer Jack Henstridge of Oromocto, New Brunswick, actually uses river-run gravel as his aggregate material in his mix. He only removes stones which do not pass between the fingers of a gloved hand, and if a few "potatoes" slip through, he removes them if they happen to get in the way of laying up log-ends. He still uses a sawdust admixture for the reasons cited above. It strikes me that a mud with a gravel base would be harder to point (pointing will be discussed later), but Jack is not as fastidious with the pointing as my wife, Jaki.

LIME, PORTLAND CEMENT, AND MASONRY CEMENT

It is important to know the differences between the various materials used as a binder in mortar mixes.

Lime. The lime we are concerned with is builder's lime, bought where cement is sold, and also called hydrated or Type S lime. This is not to be confused with agricultural lime which is used as a nutrient in soils.

Portland cement. It is from Portland cement that the mortar gets its strength, hardness, and rapid setting ability. It is a manufactured material, made by burning and refining lime, silica, alumina, iron oxide, and gypsum. The U.S. Bureau of Standards maintains rigid specifications for Portland cement, so you can be sure of the strength characteristics if you buy a bag of Type 1 (also called "regular") Portland. The standard bag is one cubic foot and weighs 94 lbs. This is the kind commonly sold. Other Types, such as 2, 3, and 4, are made for various other special requirements.

Masonry cement. This is the one where we have to be a little careful. Masonry cement is a mixture of Portland cement and lime, but the proportions will vary according to "Type." If no Type is specified, it is hard to be real sure of what proportions you've got. Generally, most common masonry cements—sometimes known erroneously as "mortar"—are made in a proportion of about two (2) parts Portland to one (1) part lime. But in some parts of the country, the common masonry cement may be somewhat weaker in terms of Portland cement. I've had good success with Type N masonry cement. Type S is also very strong. If you can't get Types S or N or don't know the Type, my advice is to stick with the Portland cement mix given below. Type 1 (or "regular") Portland cement is the same throughout the land. Even if you know that the masonry cement is Type N or S, you may still wish to do a cost comparison of the two mixes given below to find out which is cheaper. This varies quite a bit from time to time and from place to place.

Always buy the same brand and Type of cement, or you will notice variations in the color of the mixes. In the past, for example, we have found that masonry cement from Canada has been much whiter than that from New York.

While on the subject of color, using cement from our part of the world, we have found that the Portland mix sets to a lighter color than the masonry mix. But in the last few years, a new product has appeared, called "light" masonry cement. We have light beer, light potato chips, why not light masonry cement? In this case, though, the "light" refers not to strength or caloric content, but color. A mortar made from light masonry cement is very much lighter in color than that produced from the ordinary masonry cement. It is quite similar to the color of mortar which comes from the Portland mix. As cordwood masonry is light-absorbing, this is an important consideration. Under no circumstances would I advise you to repeat our mistake of buying white Portland cement. It is *very* expensive and the result is really not very much lighter in color than regular Portland or "light masonry" cement. The color of the sand and sawdust influences the color more than the Portland. Similarly, I do not advise the use of powdered dies, such as pink or black. The colored dies look artificial and it is hard to maintain consistency of color. They also add time and expense to the project. And they will darken the mix as they color it. In my view, a natural light masonry grey sets off the log-ends best. And it is the log-ends, not the mortar, which provide the unique visual interest in a cordwood wall.

RUBBER GLOVES

Stop! Do not even attempt to mix mortar—or even think about laying up a cordwood wall—without buying and religiously using a pair of good rubber mason's gloves. These gloves are cloth-lined. I've seen them black, blue, red, orange, and even green. If you don't use them, your hands may end up the same way. An acceptable alternative is a pair of Playtex or similar Heavy Duty Household Gloves. These are cloth-lined and have a good heavy rubber coating. Unacceptable are thin dish-washing gloves or surgeon's gloves. Not only do these gloves sweat like a prizefighter in the 12th round, but they rip within the first hour of use and will then be of no use whatsoever. The gloves with the heavy cloth lining sweat less and are easy to take on and off quickly. They take a day or two to get used to, but please make the effort. If you don't, I will not want to receive your correspondence. There's nothing more painful than the nasty little "cement holes" that will appear on your hands from handling mortar without rubber gloves. And the sores take forever to heal.

THE MORTAR MIX

I have deliberately refrained from discussing the actual mortar recipes to this point, not wanting to confuse the issue with a lot of inferior mixes. The two mixes I will describe are roughly equivalent chemically and in terms of strength, assuming a good quality (Types N or S) masonry cement. The masonry mix will be slightly darker, perhaps, unless light masonry cement is used.

Most of the buildings at Earthwood use the masonry cement mix of:

9 sand, 3 soaked sawdust, 3 masonry, 2 lime.

The proportions are equal parts by volume, not weight. Using an ordinary long-handled spade, with equally rounded shovelfuls, these proportions will yield a good workable amount of material in a contractor's wheelbarrow. I always keep two spades on hand, one for the dry goods and one for the soaked sawdust or wet sand. This way, a shovel which has just been in the wet sawdust barrel does not get caked up with the cement or lime.

To greatly reduce mixing time, I stagger the ingredients into the wheelbarrow according to the following cadence:

3 sand—1 sawdust—1 masonry—1 lime
3 sand—1 sawdust—1 masonry—1 lime
3 sand—1 sawdust—1 masonry.

Note that there is no lime in the third line. The two shovels of lime are buried towards the middle of the mix. If anyone starts talking to you while you are mixing, just continue your count out loud until they shut up. They'll soon get the idea. If your sand is quite coarse, you may find that an extra shovel of lime—in the third line of the cadence, of course—

2-2. (left) Mix the dry ingredients with a hoe. 2-3. (right) Jaki adds water in a crater, then continues mixing with a hoe.

will make the mortar more plastic and, therefore, more workable.

Dry-mix the batch with a hoe. Always pull the ingredients towards you, using small chopping motions just before each pull. Pulling is much easier than pushing. When all the stuff is at one end, take a small rest while you walk to the other end of the barrow. Now pull everything towards that end. After three or four trips, the "dry-mix" is complete.

How much water should you add? Good question, but there is no exact answer, because the amount of water will depend on how wet the sand is, and how much water you're pulling up with each trip to the sawdust barrel. Listen:

Make a little crater in the center of the mix and add a splash of water. Let it sink in for a few seconds and then start pulling the dry goods into the crater. Soon, you will be back to pulling from one end of the barrow to the other, as you did during the dry mix. Now you'll start to appreciate those little rests. Remember to get your hoe down to the corners of the wheelbarrow to pull up the dry goods lurking there. Keep adding water and mix it in thoroughly. The final mix should be quite damp, but definitely not "soupy." The mortar should be thick enough to support the log-ends without slumping and to hold its shape when placed against an adjacent log-end or up against the straight side of the wheelbarrow. Try the "snowball test." Make a ball of the mortar and toss it up in the air about two feet and catch it in your rubber gloves. It should not crumble from being too dry. It should not splash or flatten out from being too wet. From two feet up, the ball should give just a little

when you catch it. This is what I mean by "plastic."

A final note of warning: When the mortar is very close to being just right, be very cautious with the water. Add just a tiny splash, little more than a blessing. At this point, a very small amount of water can suddenly make the whole mix soupy. If this happens, you will have to add more dry goods in the same proportion as the mix you are using, just enough to stiffen it up as needed. Mix the new stuff in thoroughly.

The Portland mix is similar. We have used it in a 16-ft diameter shed at Earthwood. In some ways, I think I prefer it to the masonry mix. It is strong and

2-4. The "snowball test."

hard but still has the nonshrink characteristic. It is lighter in color, although the use of "light" masonry cement will negate this advantage. Again, see what is available in your area and cost it out. The Portland mix is:

9 sand, 3 soaked sawdust, 3 lime, 2 Portland.

The cadence for placing ingredients in the barrow is:

3 sand—1 sawdust—1 lime—1 Portland
3 sand—1 sawdust—1 lime—1 Portland
3 sand—1 sawdust—1 lime.

Proceed as per the instructions above for making the masonry cement mix.

The mixes given above have given us very good results, but they are by no means the only mixes possible. At Log End Sauna, built a year before Earthwood, we had good luck with a mix of 7 sand, 5 sawdust, 3 masonry, 1 lime. These walls have weathered well, which is to say they have weathered very little indeed. There is noticably more sawdust in the finish, and the mud is slightly more difficult to point smoothly. It probably has more insulation value, however. The equivalent Portland mix would be: 7 sand, 5 sawdust, 2 Portland, 2 lime.

Cordwood author and builder Richard Flatau likes a mix of: 3 sand, 3 sawdust, 1 Portland, 1 lime.[8] This mix is very similar to the one given just above, but is even heavier with sawdust. Richard's house, called "Flatau's Plateau," is a post and beam building with cordwood infilling. If the cordwood wall is to be load-bearing, he recommends cutting back to 1 part sawdust, which would make a strong mix (3:1:1:1) almost identical in proportion to our Earthwood mix of 9:3:3:2.

Jack Henstridge, who has helped put hundreds of people in affordable cordwood houses thanks to books, slideshows, and seminars, has, in the past few years, turned to using what he calls a "Concrete Mortar." The aggregate, instead of sand, is "pit run" gravel, which is, he says, "basically the same material you'd use in a standard concrete mix." The proportions of Jack's mix are: 3 pit run gravel, 2 sawdust, 1 Portland, 1 masonry cement. This is a strong mix and, says Cordwood Jack, "leaves nothing to be desired."[9] Obviously, a mix with an aggregate of pit run gravel will not be as smooth as one where sand is the main component. But Jack's view of pointing is a little different from mine:

As the wall builds up, smooth out the mortar between the wood blocks or even rout it out with your fingers so the ends of the blocks actually stick out a little. (Looks real nice.) Do not worry about getting it smooth or even; it can be done, but if you want smooth walls, this is not the method to use. The idea is a rugged wall and that is what you are going to get.[10]

CEMENT MIXER VS. WHEELBARROW

An electric or gas-powered cement mixer is particularly useful in work situations where there are at least two masons and one designated laborer. With a mixer, the proportions to be used are the same as given above. Jack Henstridge recommends the use of measuring containers, such as plastic buckets, to assure uniformity of proportions.

In the case of a person working alone, or even a couple, I feel quite strongly that mixing in a wheelbarrow has several advantages over a powered mixer:

(1) There is no great saving in time with a mixer. With a little practice, the 9:3:3:2 mixes described above can be made in 9 or 10 minutes. The actual mixing time might be a little less with the mixer, but the mix still has to be dumped into a barrow for transport, so twice as much equipment needs to be cleaned. Add to the time the downtime fetching gas, transporting the mixer, battling with the infernal frustration engine ...

(2) It is easier to "fine-tune" the mix with a barrow. The mud is less likely to go "soupy." It is easier to test the slump or plasticity.

(3) There is a slight risk of injury with the mixer. Most of this can be eliminated by keeping children away from the mixer and never, *ever* putting a shovel, hoe or other tool in close proximity to the rotating drum or paddles.

(4) Mixers are noisy and vexatious to the spirit.

The exception to my commentary would be if the builders have some physical condition which makes hand-mixing too strenuous, such as heart trouble. Having said that, please don't get a heart attack yanking on a pull cord to start your mixer or generator. Find an easier way or hire a high school football player to mix the mud. . .

Hand-mixing is the most physically demanding part of cordwood masonry; laying up the wall is a piece of cake by comparison. It can keep you physically fit, but you'll always be glad when each batch passes the snowball test and is ready to use.

TOOLS FOR MUD

For mixing, hauling, and keeping mud close by you will need a wheelbarrow, rubber gloves, a hoe, two shovels (long-handled spades are best), a couple of five-gallon buckets, a mortar board, and a stiff brush (plastic lasts longest) for cleaning tools. The mortar board can be a fibreglass one made for the purpose, or you can improvise. I use the tops of 55-gallon drums, and some very handy long aluminum trays I bought at an auction. Porous material, such as plywood, is not so good as it robs moisture from the mud. With two partners working together, it is a good idea to double up on some of the tools. Two wheelbarrows, two hoes, two mortar boards, for example. If there is running water on site, a hose with spray attachment would be a wonderful luxury.

For laying down the mortar, rubber gloves work best, but a mason's trowel and a small pointing trowel can be handy, such as when placing mortar against posts and up to the underside of beams.

CLEAN YOUR TOOLS!

Whether you mix in a barrow or a powered mixer, always clean the mixing vessel completely at the end of each day's work. I even clean my barrows at lunch, and despite twelve years' of hard use, both of my wheelbarrows are almost as good as new. The same respect must be given to shovels, hoes, mortar pans, levels, trowels and pointing knives. Do not let cement harden on any of your steel tools.

Of particular importance is to keep your gloves free of hardening mortar. Wash them thoroughly at the end of each day in a bucket of water while they are on your hands, being careful not to get water inside. (If you do, turn 'em inside out to dry.) Such care is not fastidious. It is economic. Good gloves cost about $5 a pair and they will perish quickly without care.

KIND OF WOOD

When workshop students ask what kind of wood should be used with cordwood masonry, my stan-dard answer is, "Use what you've got." Almost any wood can be used, as long as it is sound. Do not use spongy or punky wood, or unstripped birches, as the bark will trap the wood's moisture and the log-ends may turn punky after they are laid up. Actually, it's best to strip any wood to be used with cordwood masonry, as insects just love the area between the bark and the outer wood layers. Plus, stripped wood dries a whole lot faster. Wood can be stripped of its bark best in the spring, when rising sap loosens the bark from the epidermal layers.

But let's say you've got a choice. Several different kinds of wood are available locally, perhaps even on your property. What are some of the considerations in making a choice?

CORDWOOD AS INSULATION

Wood's insulative qualities are just about proportional to its "lightness of being." Light, airy woods like northern white cedar, larch, basswood, and even poplar (quaking aspen) will have good insulative qual-ities (high "R" values) compared with the heavier and denser woods like oak, hard maple, red heart cedar.

Insulation values of wood are generally given through side grain. Unfortunately for cordwood builders, heat loss through wood is much greater along end grain. (On the positive side, it is this same breathability that so greatly prolongs the life of a cordwood wall.) I have seen several different esti-mates of a coefficient—let's call it "C"—by which side-grain R-values can be converted to end-grain values. The most pessimistic value for C that I've seen was in a thesis presented to the Department of Industrial Education, Brigham Young University, by Gregory M. Kortman in April, 1989. Using figures derived from an article appearing in a 1952 U.S. Department of Agriculture handbook called *Preservative Treatment of Wood by Pressure Methods*, and distilled through a textbook called *Flow in Wood* by John F. Siau (Syracuse University Press, 1971), Kortman hugs tightly to a value of C of 0.40. In other words, the R-value of a wood along end grain is con-sidered to be only 40 percent that of its value through side grain. Several other cordwood writers have published articles falsely assuming a 1.00 (100 percent) value for C. I have come across other values and estimates from various sources.

It is likely that the true value of C varies with the type of wood used. At Earthwood, for point of discussion (and in the following chart), we use a C-value of 0.66, or about ⅔. The important point is that heat loss is greater through end grain than through side grain. It is also important to know that the insulation value for the mortar matrix in a 16" cordwood wall is actually greater than that of the wood portion. (6" of sawdust at R3.5/inch = R21. Add about R2 for each mortar joint [conservative] and you have an R25 mortar matrix.)

The following chart is by no means intended to be complete, as only a few woods are included, nor is it touted as the last word in accuracy. The waters are still murky in this area. It does, however, give a good idea of the kinds of consideration that have to weighed up when selecting which wood to use.

Wood Species	R-value/inch (side grain)	R-value/inch (end grain, as log-ends.)
White cedar	1.50	1.00
Poplar, Hemlock	1.20	0.80
Douglas fir	1.14	0.76
Oak, Maple and other dense hardwoods	0.90	0.60
Red heart cedar	0.90	0.60

Now, let's play with numbers for a bit. Let's compare the R-values of three different 16" cordwood walls, built respectively out of white cedar, poplar, and oak. The cordwood portion of the cedar wall will have a value of R16 (R1.00 × 16"). But a cordwood wall is only about 60 percent cordwood; the rest is mortar. (Is the reader surprised that 40 percent of a cordwood wall's area is mortar? The author was, too, but precise measurements and calculations on two different panels at Earthwood each returned the same result.) The mortared portion of the wall averages about R25, as already noted above. I'll spare the algebra, but the average R-value of the 16" cedar wall is R19.60.

With the poplar (or hemlock) wall, the wood portion is worth R12.8 (R0.80 × 16") and averaged with the R25 mortar portion over 40 percent of the wall, the wall's total average is R17.68

With an oak (or other dense hardwood) wall, the log-end part is worth R9.6 (R0.60 × 16"). Averaged with the mortar matrix, we come up with R15.76.

Greg Kortman, using a coefficient of 0.40 for C

would come up with much lesser values. Jack Henstridge, in a 1984 update to his *Building the Cordwood Home*, figures an average thermal resistance of R26.4 on a 16" wall with low-density woods at an uncorrected value of R1.5/inch and walls which are 80 percent cordwood by surface area (high).

Tests performed by the University of Manitoba point to an R-value of about R1 per inch. There, a 24" thick cordwood wall prototype was built using poplar and spruce, with an insulated space between the inner and outer mortar joints. The insulation value of this prototype was rated at R24.[11]

CORDWOOD MASONRY AS THERMAL MASS

But insulation is only half the story when it comes to evaluating a cordwood wall's thermal performance. Better than any other method of building that I can think of, cordwood masonry combines insulation with exceptional thermal mass, the ability of a material to store heat. A stone wall, like a Scottish cottage (or castle), has great mass, but poor insulation. The modern super-insulated house has superb insulation but practically no thermal mass.

In a cordwood wall, the log-ends have both mass and insulation characteristics. Oak would have more thermal mass than white cedar, for example, but less insulative value, as we have seen. It is reasonable to think of the insulative values of various kinds as being inversely proportional to their values as ther-

2-5. How a cordwood wall works.

mal mass. The mortar matrix also has qualities of both insulation and thermal mass, but there is a significant difference. Most of the insulation value is to be found in the filled cavity between the inner and outer mortar joints. Most of the thermal mass is to be found in the mortar itself. Wonderfully, there is a thermal mass—literally tons—on each side of the insulation. How is this to our benefit? Well, a cordwood wall works like this, as shown in 2-5:

In the winter, the internal heat source, such as the wood stove shown, radiates heat to the cordwood walls, "charging up" the inner mortar joint with heat. That heat does not readily transfer to the outside because of the heavily insulated space inside the mortar matrix. At no point in a cordwood wall does the heat wick out of the house directly through the dense mortar.

There are no "energy nosebleeds." The heat is "stored," ready to be given back into the room as the room temperature tries to drop. The log-ends themselves also store heat, but some is also wicked out. There is a gradual decrease in temperature through the log-end portion. In the mortar matrix portion, the inner mortar joint is warm, the outer mortar joint is cold. The two extremes are protected from each other by the insulation.

So what good is the cold outer portion of the mortar matrix? Not much, in the winter. But in the summer (or in the south), it is of tremendous value! The hot direct rays of the sun (or maybe just the heat of the day) cause the outer mortar joint to heat up. But, again, this heat does not readily conduct to the inside because of the insulation cavity. At night, the day's heat gain is lost back into the atmosphere by radiant cooling. The inside of the wall is not affected by extremes. Cordwood houses are naturally cool in the summer.

The proper placement of insulation is on the cold side of the thermal mass, so that the mass can store heat. This is why we place Styrofoam® on the exterior of block walls in earth-sheltered housing, as will be seen in Part Two. The walls at Earthwood are 40 percent earth-sheltered, with 60 percent cordwood (and windows and doors) above grade. We heat the two-storey 2200 sq ft house with only 3¼ full cords of firewood each year, equivalent to 432 gallons of fuel oil. This takes care of most of our winter cooking and hot water heating, too, not bad for an area with nearly 9000 degree days, just 60 miles from Montreal.

ROT RESISTANCE

A common concern amongst prospective cordwood builders is that the wood might rot in the wall, so they want to know which woods have the better resistance to decay. This is not really as important a

Grouping of some domestic woods according to approximate relative heartwood decay resistance		
Resistant or very resistant	Moderately resistant	Slightly or nonresistant
Bald cypress (old growth)	Bald cypress (young growth)	Alder
Catalpa		Ashes
Cedars	Douglas-fir	Aspens
Cherry, black	Honeylocust	Basswood
Chestnut	Larch,	Beech
Cypress, Arizona	western	Birches
Junipers	Oak, swamp	Buckeye
Locust, black[1]	chestnut	Butternut
Mesquite	Pine, eastern	Cottonwood
Mulberry, red[1]	white	Elms
Oak:	Southern Pine:	Hackberry
Bur	Longleaf	Hemlocks
Chestnut	Slash	Hickories
Gambel	Tamarack	Magnolia
Oregon white		Maples
Post		Oak (red and
White species)		black
Osage orange[1]		
		Pines
		(other
Redwood		than long-
Sassafras		leaf, slash,
Walnut, black		and eastern
Yew, Pacific[1]		white)
		Poplars
		Spruces
		Sweetgum
		True firs
		(western
		and eastern)
		Willows
		Yellow-poplar

[1] These woods have exceptionally high decay resistance.

concern as one might think, if three basic rules are followed:

(1) Use sound wood to begin with, no matter what the variety. There should be no deterioration in evidence. Bark should be removed so as not to trap moisture or harbor insects.

(2) Keep the bottom course of wood off the ground, by at least four inches (4"). To stop vegetation from growing up next to a cordwood wall, I always make a skirt of black plastic covered with two inches of crushed stone. Water dripping from the eaves does not erode the earth around the building or cause mud to splash on the cordwood wall.

(3) Use a good overhang on the roof. I would consider a foot to be bare minimum. At Earthwood, the overhang of the 16-sided roof varies from 16" to 22".

Having said that, I offer the chart on the previous page. It is reproduced with permission from *The Encyclopedia of Wood* (Sterling, 1989), an excellent work which contains detailed commentary on practically any commonly available wood.

Do not be put off if the wood you have in mind is in the third column. I have seen successful cordwood houses made with red pine and even white birch (providing the bark is removed very soon after felling the tree.) Hemlock has been used with some success, the main problem with it being that it shrinks more than practically any other wood. The hardwoods like maple, oak, beech, and elm also tend to shrink quite a bit.

SHRINKAGE AND EXPANSION

All woods swell when they take on moisture and shrink when they dry out. The chart on the following two pages, from *The Encyclopedia of Wood*, gives the shrinkage from green to oven-dry moisture content We are particularly concerned with radial and tangential shrinkage, the kinds that make for loose log-ends in a wall.

Note that northern white cedar is tied with young growth redwood for having the least radial shrinkage. Radial shrinkage causes the radial check lines in an unsplit log-end. Usually one large "primary check" will win out and open wider, its brothers remaining as hairlines. Only eastern red cedar has (slightly) less tangential shrinkage than the northern

white variety. Tangential shrinkage is what causes the log-end to actually loosen up in the wall. Think of this shrinkage as the kind experienced when a balloon loses its air.

Shrinkage is a more common—but less serious—problem than expansion. By using dry white cedar and old cedar fence rails, we have managed to escape with very little shrinkage in our cordwood walls, not enough to cause draft or to necessitate any kind of cosmetic cure. The exception, at Earthwood, was seven large round hardwood log-ends which shrunk a lot. These logs—six elm and one 19"-diameter beech which had been drying already for three years—were used as design features. They shrunk so much that you could see daylight all around them. The best solution for just a few bad eggs like this is to caulk around them with clear silicone or silicone-based caulking. Others have used white, grey, or brown caulk, but this is not a success, esthetically. It looks artificial. The clear caulk is virtually unnoticeable.

As silicone caulking—or even the more economic silicone-based caulking, such as Red Devil Lifetime®—is fairly expensive, it might not be a good choice where all the log-ends of the house have shrunk. In this case, make sure the log-ends have done all their shrinking—wait two years to be sure—and then apply Thoroseal®or equal masonry sealer to the mortar joints with a one-inch brush. Thoroseal® (made by Thoro Corporation) is a Portland cement–based powder which is mixed with water to a thick paste. It comes in white or grey. The paste should have enough body to cover over or leach into the cracks around the log-ends. If the masonry sealer is drying out too rapidly on the old mortar joints and flaking off, apply a coat of Acryl-60 (also made by Thoro Corporation) or an equal bonding agent to the old mortar prior to the Thoroseal®. Acryl-60 is a thin milky-white substance that creates an almost miraculous bond with cement-based products.

Another option for taking care of wood shrinkage has been used successfully at the homes of Richard Flatau in Wisconsin and Harold Johnson in New York, featured in Chapter Six. Both builders, independently, came up with the use of Perma Chink (Perma Chink Systems, Inc., 1605 Prosser Rd., Knoxville, TN 37914), a material developed as a permanent cure for expansion and contraction problems associated with horizontal log chinking. In an

SHRINKAGE VALUES OF DOMESTIC WOODS

Species	Radial	Tangential	Volumetric	Species	Radial	Tangential	Volumetric
	Shrinkage from green to ovendry moisture content[1]				Shrinkage from green to ovendry moisture content[1]		
	Percent				*Percent*		

HARDWOODS

Species	Radial	Tangential	Volumetric	Species	Radial	Tangential	Volumetric
Alder, red	4.4	7.3	12.6	Hickory, True:			
Ash:				Mockernut	7.7	11.0	17.8
Black	5.0	7.8	15.2	Pignut	7.2	11.5	17.9
Blue	3.9	6.5	11.7	Shagbark	7.0	10.5	16.7
Green	4.6	7.1	12.5	Shellbark	7.6	12.6	19.2
Oregon	4.1	8.1	13.2	Holly, American	4.8	9.9	16.9
Pumpkin	3.7	6.3	12.0	Honeylocust	4.2	6.6	10.8
White	4.9	7.8	13.3	Locust, black	4.6	7.2	10.2
Aspen:				Madrone, Pacific	5.6	12.4	18.1
Bigtooth	3.3	7.9	11.8	Magnolia:			
Quaking	3.5	6.7	11.5	Cucumber tree	5.2	8.8	13.6
Basswood,				Southern	5.4	6.6	12.3
American	6.6	9.3	15.8	Sweetbay	4.7	8.3	12.9
Beech, American	5.5	11.9	17.2	Maple:			
Birch:				Bigleaf	3.7	7.1	11.6
Alaska paper	6.5	9.9	16.7	Black	4.8	9.3	14.0
Gray	5.2	—	14.7	Red	4.0	8.2	12.6
Paper	6.3	8.6	16.2	Silver	3.0	7.2	12.0
River	4.7	9.2	13.5	Striped	3.2	8.6	12.3
Sweet	6.5	9.0	15.6	Sugar	4.8	9.9	14.7
Yellow	7.3	9.5	16.8	Oak, red:			
Buckeye, yellow	3.6	8.1	12.5	Black	4.4	11.1	15.1
Butternut	3.4	6.4	10.6	Laurel	4.0	9.9	19.0
Cherry, black	3.7	7.1	11.5	Northern red	4.0	8.6	13.7
Chestnut,				Pin	4.3	9.5	14.5
American	3.4	6.7	11.6	Scarlet	4.4	10.8	14.7
Cottonwood:				Southern red	4.7	11.3	16.1
Balsam poplar	3.0	7.1	10.5	Water	4.4	9.8	16.1
Black	3.6	8.6	12.4	Willow	5.0	9.6	18.9
Eastern	3.9	9.2	13.9	Oak, white:			
Elm:				Bur	4.4	8.8	12.7
American	4.2	9.5	14.6	Chestnut	5.3	10.8	16.4
Cedar	4.7	10.2	15.4	Live	6.6	9.5	14.7
Rock	4.8	8.1	14.9	Overcup	5.3	12.7	16.0
Slippery	4.9	8.9	13.8	Post	5.4	9.8	16.2
Winged	5.3	11.6	17.7	Swamp chestnut	5.2	10.8	16.4
Hackberry	4.8	8.9	13.8	White	5.6	10.5	16.3
Hickory, Pecan	4.9	8.9	13.6				
Persimmon				Tupelo:			
common	7.9	11.2	19.1	Black	5.1	8.7	14.4
Sassafras	4.0	6.2	10.3	Water	4.2	7.6	12.5

Table continued

SHRINKAGE VALUES OF DOMESTIC WOODS continued

Species	Shrinkage from green to ovendry moisture content[1]			Species	Shrinkage from green to ovendry moisture content[1]		
	Radial	Tangential *Percent*	Volumetric		Radial	Tangential *Percent*	Volumetric

HARDWOODS continued

Species	Radial	Tangential	Volumetric	Species	Radial	Tangential	Volumetric
Sweetgum	5.3	10.2	15.8	Walnut, black	5.5	7.8	12.8
Sycamore				Willow, black	3.3	8.7	13.9
American	5.0	8.4	14.1	Yellow-poplar	4.6	8.2	12.7
Tanoak	4.9	11.7	17.3				

SOFTWOODS

Species	Radial	Tangential	Volumetric	Species	Radial	Tangential	Volumetric
Baldcypress	3.8	6.2	10.5	Hemlock (con.):			
Cedar:				Western	4.2	7.8	12.4
Alaska-	2.8	6.0	9.2	Larch, western	4.5	9.1	14.0
Atlantic white-	2.9	5.4	8.8	Pine:			
Easter				Eastern white	2.1	6.1	8.2
redcedar	3.1	4.7	7.8	Jack	3.7	6.6	10.3
Incense-	3.3	5.2	7.7	Loblolly	4.8	7.4	12.3
Northern				Lodgepole	4.3	6.7	11.1
white-	2.2	4.9	7.2	Longleaf	5.1	7.5	12.2
Port-Orford-	4.6	6.9	10.1	Pitch	4.0	7.1	10.9
Western				Pond	5.1	7.1	11.2
redcedar	2.4	5.0	6.8	Ponderosa	3.9	6.2	9.7
Douglas-fir:[2]				Red	3.8	7.2	11.3
Coast	4.8	7.6	12.4	Shortleaf	4.6	7.7	12.3
Interior north	3.8	6.9	10.7	Slash	5.4	7.6	12.1
Interior west	4.8	7.5	11.8	Sugar	2.9	5.6	7.9
Fir:				Virginia	4.2	7.2	11.9
Balsam	2.9	6.9	11.2	Western white	4.1	7.4	11.8
California red	4.5	7.9	11.4	Redwood:			
Grand	3.4	7.5	11.0	Old-growth	2.6	4.4	6.8
Noble	4.3	8.3	12.4	Young-growth	2.2	4.9	7.0
Pacific silver	4.4	9.2	13.0	Spruce:			
Subalpine	2.6	7.4	9.4	Black	4.1	6.8	11.3
White	3.3	7.0	9.8	Engelmann	3.8	7.1	11.0
Hemlock:				Red	3.8	7.8	11.8
Eastern	3.0	6.8	9.7	Sitka	4.3	7.5	11.5
Mountain	4.4	7.1	11.1	Tamarack	3.7	7.4	13.6

1 Expressed as a percentage of the green dimension. 2 Coast Douglas-fir is defined as Douglas-fir growing in the States of Oregon and Washington west of the summit of the Cascade Mountains Interior West includes the State of Califomia and all counties in Oregon and Washington east of but adjacent to the Cascade summit. Interior North includes the remainder of Oregon and Washington and the States of Idaho, Montana, and Wyoming.

itself), they will expand and cause the mortar joints to crack. I'll return to this subject real soon under "Drying the Wood."

KIND OF WOOD, CONTINUED

Now that we have discussed some of the qualities of wood, we can better judge the species available to determine order of preference.

The best wood, if available, is northern white cedar. It has excellent insulative value, is easy to work with, looks great, weathers well, has very high rot resistance, and shrinks very little. It will not cause severe expansion problems, even when "popcorn dry." Earthwood is mostly old split cedar fence rails, with some cedar rounds thrown in. We used some hardwoods as special design features and near the top of the wall when we began to run low on cedar.

Ron Farrell of South Dayton, New York, built a beautiful Earthwood-type house of larch. Even though larch is in the second column of "moderately resistant" woods, there is no doubt that Ron's wood will last 100 years or forever, whichever comes first. The larch has the same light airy quality as white cedar.

Poplar (quaking aspen) is considered by many to be a junk wood, of little value for anything. The wood is much maligned, in my opinion. We have an excellent Adirondack deck table made from "popple," and have used it successfully as sauna benches because of its low conductivity of heat. We also use it for stovewood in the spring and fall. Many successful cordwood houses have been built with poplar throughout North America, including Harold Johnson's.

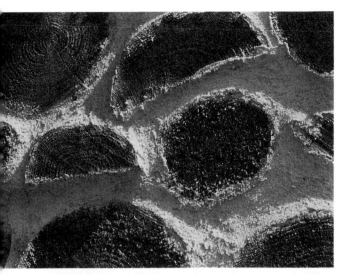

2-6. Perma Chink seals the wood shrinkage gaps at Harold Johnson's house.

update to his book *Cordwood Construction: A Log End View*, Richard says:

> Perma Chink is an "acrylic latex co-polymer based sealant" which can be thinly spread (with a stiff brush) over the existing mortar, to create a durable, resilient surface. Perma Chink adheres to the wood and mortar and doesn't crack or separate with the onslaught of the seasons. It can be used on the interior or the exterior and comes in five mortar-like colors. Perma Chink provides a viable remedy for logs that have loosened or mortar that has cracked.[12]

The material is a little expensive at about $75 per five gallon bucket, but Harold Johnson was able to do the entire exterior of his home with three buckets. I visited the home in September of 1991, after the Perma Chink had been on for about two years. It looked great and was still flexible.

Wood shrinkage is an irritation, but can be attended to a year or two after the house is built as described above. Wood expansion, however, can be a structural problem. Wood expansion can actually break up your walls. I know. It happened to me, as will be related in Part Two, Chapter 12. Luckily, it is a rare problem, and occurs when dense hardwood log-ends are laid up *too* dry. If they take on moisture from any source (high humidity, driving rain, water collecting on a slab foundation, the wet mortar

BARKING THE WOOD

Jack Henstridge and I disagree with those who advocate the use of unbarked log-ends. We have several objections. First, the bark greatly inhibits the proper drying of the wood; second, insects love to get in between the bark and the epidermal layers of the wood; and third, although the rough bark might adhere to the mortar mix very well, as other writers in the field have pointed out, the eventual loosening of the bark-to-wood bond renders the bark-to-mortar bond superfluous. Jack says,

> An indication of the necessity of [barking] can be shown by using the town of St. Quentin, New Brunswick, as an example. In the early twenties the community was burned down and later rebuilt using the burned trees. The first houses built used relatively "green" wood with the bark still on it. These houses did not last. The homes built later (approximately three to four years), after the wood had dried, shrunk, and the bark had fallen off are still standing to day and as solid as the day they were built This was probably the last time (cordwood) construction was used on an extensive scale in North America.[13]

At any time of the year it is good to bark the tree soon after felling, and the same day if possible. The easiest time of the year to bark is spring, when the rising sap loosens the tight bond between bark and wood. In April, we could pull the bark off our cedar logs with our fingers. Sabre, our German shepherd, had a great time yanking 8' strips away from the log. That dog could really bark. The worst time of the year for barking is late autumn, when the sap has run back into the ground to protect the tree from freezing. The bark is then extremely reluctant to let go of the log.

As stated, we barked in the spring, so it was an easy job with an axe, a trowel, and a dog. We would force the blade of the axe or trowel under the bark at a tangent to the log, working the blade right and left. Because the wood was freshly cut, the bark pulled away like a banana peel. Later in the summer, Jaki and I helped friends bark cedar logs which had been cut a few weeks earlier. It was a completely different job. Armed with peeling spuds made from truck springs (a big improvement over a trowel and an axe), we were still lucky to complete one 14' log in an hour. The moral: peel fresh. Though peeling logs was not a problem for us, it can be a troublesome task. Christian Bruyere's *In Harmony with Nature* (Sterling, 1975) includes an illustration by Robert Inwood showing clearly one method of making a handy peeling spud (2-7). Though I have not tried this tool myself, it makes more sense than the short-handled spuds we used on our friends' logs. The longer handle should give the user a vastly superior leverage.

Tom Hodges, in a short article in *The Mother Earth News (July, 1976)*, has another idea. He says,

> My debarker is simply a garden hoe with the blade straightened out. To make one just take an old hoe, heat its "neck" until the metal is malleable, and bend the blade back until it forms a 165° angle with the handle. Sharpen the business end and presto! You've got a tool that's guaranteed to make easy work of any bark-stripping job. To use the spud, just anchor or wedge your (log) so it won't move, stand over it, and dig in. With a little practice, you'll soon be able to peel off two- to three-foot strips with one swipe.[14]

After the wood is barked, it should be cut into log-end size. A tractor saw (buzz saw) is easier than a chainsaw for this job. Two different esthetic effects can be obtained by splitting or not splitting the log-ends. There are examples of both styles in this book. The primary advantage of leaving the log-ends round is that fewer will be needed to build the wall, so there is less handling of materials. Also, some people just like the way they look.

2-7 Making a peeling spud out of an old shovel.

SPLIT WOOD

Split cordwood has several advantages over the rounds. First, splitting the wood greatly accelerates the drying process. Drying of wood takes place primarily through end grain and is extremely slow through the epidermal (outer) layers of the wood. Splitting the wood exposes more drying surface to the sun and air, especially in the heartwood area.

Second, split wood enables the builder to keep a

constant mortar joint width, which facilitates pointing and strengthens the wall. Figure 2-8a shows rounds laid up in the most efficient pattern, a honeycomb configuration. You can see that there are small mortar joints where the ends are tangent and a large triangular area of mortar between three adjacent ends. Figure 2-8b shows a section of split ends. The mortar joint is held constant, making it easier to use the pointing knife. Clearly, example 2-8b is the stronger of the two. Try stacking split firewood in one rank and the same quantity of rounds in another, and then compare their weight-supporting abilities.

Another source of strength comes from the ease

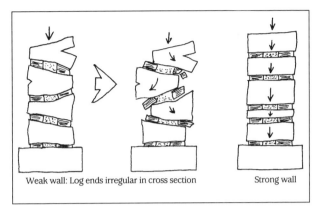

Weak wall: Log ends irregular in cross section   Strong wall

2-9. Weak and strong cordwood walls.

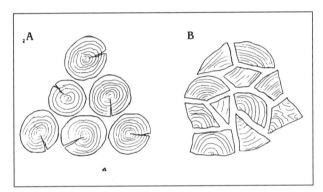

2-8. A wall of split ends is easier to point.

of pointing the randomly shaped split ends. The joint can be tightened by applying pressure with the pointing knife, more difficult to do with the different widths of the mortar joints in a round log-end panel. This pressure maximizes the bond between wood and mortar.

Jaki points out that splitting wood also eliminates the need for stuffing that big primary check, an often necessary task with large rounds.

It is important to split the wood randomly—not all quarter logs, for example. A large selection of different sizes and shapes enables the builder to easily find the piece needed to continue the wall. Rounds and split ends can be combined in a wall, of course, and usually are to some degree.

Whether you choose round or split wood, avoid tapered or irregular log-ends, such as occur near branch joins and near the foot of a tree. The problem with irregular log-ends is illustrated in 2-9.

I feel that the split wood is easier to lay properly

and to point. However, we let students at Earthwood Building School try both styles. If you can do well with different-sized rounds, you'll have no trouble with a variety of splits.

Don't let my commentary on strength put you off using rounds. "Mushwood," our summer cottage, has a round first story composed entirely of round log-ends. It supports a geodesic dome on the second story. See Chapter 6. Rohan's building at Earthwood is a 16' diameter round shed supporting an earth roof. It, too, is built entirely of rounds. Builders using stackwall corners very often use full rounds between the corners, as per Malcolm Miller in Fredericton, New Brunswick, and Ken Campbell of Idaho Springs, Colorado. So, while the carefully built wall of split ends may be a little bit stronger on compression, the walls built with rounds also exceed any reasonable strength requirements and have stood the test of time.

DRYING THE WOOD

Whether or not the wood is split, it should be stacked for drying as early as possible. Log-ends won't dry if they are piled like stones. There are many theories about which direction firewood ranks should run for maximum drying. Some say north and south, others east and west. My advice is to choose a site out of the shadows and where there is good air movement. The top of a grassy knoll would be good. My own theory is that the line of the rank should be at right angles to the line of the prevailing

2-10. Pallets used to support cordwood for drying.

2-11. Built-up corners support the end of this rank of wood.

wind, maximizing the wind effect in drying. The rank of wood should be off the ground, supported on logs or pallets. Figures 2-10 and 2-11 show different ways of supporting the end of a rank of cordwood. Wood should be dried in single ranks for best air circulation.

Finally, the top of the rank—but not the sides—should be covered to shed water.

A drying method which Jaki and I adopted at Log End Cottage worked very well for us. Late in the summer, after our post-and-beam framework was completed, we rented a portable circular saw and belatedly got around to cutting our barked cedar into the 9" ends we had decided to use. Two friends helped. We set up an assembly line where Jaki would place the log on the movable table of the saw, I would cut the 9" log-end, Joe would wheel them to the cottage, and his wife, Pat, would stack 'em up without mortar. In two days, all the panels were filled in with log-ends. We covered the exterior with cheap ½" insulation board, stuffed the largest unfilled spaces with leftover fibreglass from the roof, and moved in in December.

The woodstoves dried the log-ends for a good six months before the weather and the garden allowed us to get back to work on the cottage. Considerable checking had appeared in the end grain (see Fig. 2-

12). Beginning in June, we dismantled and rebuilt (with mortar) each panel, one at a time. A panel is a section between posts, usually about 5' by 6'.

The drying method described above involves a lot of handling of materials. If, like us, you're trying to live in the house at the same time, tempers can run short as the living conditions become more and more chaotic. But it works.

2-12. These log-ends were stacked to dry through a winter. The author points out a check that formed during the drying process. The panel on the left is still covered with insulation board. The panel on the right is complete.

HOW LONG SHOULD THE WOOD DRY?

My advice on how long to dry the wood comes from a baptism of fire: a trying experience with over-dry woods which swelled and broke up the walls which were to be built below grade at Earthwood. For five years, Jaki and I endeavored to find and use the driest log-ends possible in our walls. And the care we took yielded good results. The only potential problem with the wood that we knew about was shrinkage, and, by golly, we avoided that pretty well. At the same time, we were only using very dry northern white cedar, which is very nearly immune from the other problem, the potentially more serious problem, the problem we didn't know about: wood expansion. So we were lucky for five years and jumped to the seemingly logical conclusion that log-ends should be absolutely dry before laying them up in the wall.

The error in our "logic" comes from the fact that wood is a living, breathing, nonstatic material. Most importantly, each species is different. We "lucked out" with cedar. We learned the hard way that dense wood—such as maple, cherry, beech, and other hardwoods—swells if it is very dry and takes on moisture. The detailed story is recounted in Part Two, Chapter 12. I have had to reevaluate my thinking on how long to dry the wood.

With the northern white cedar, even if popcorn-dry, there is very little swelling even if the wood takes on moisture. We've had no problem in 16 years. It is as if the dry white cedar is airy enough so that the wood fibres can expand within the confines of the log-end itself, without expanding the piece. The densely packed fibres of most hardwoods, on the other hand, have no place to go when they take on moisture except outwards. Another way of looking at it is that woods which shrink the most from green to dry may also be those which will expand the most from dry to wet, as in, "What goes up must come down." Wood being wood, though, this may not be true in all cases.

In general, if the wood being considered is low on shrinkage, dry it as long as you can afford to wait, although I think most of the drying will occur in one full year. With woods that shrink a lot (and therefore have the potential to swell a lot), I'd strip, split, and

stack the wood in open ranks (tops covered) for just 6 to 10 weeks prior to construction, depending on drying conditions. Yes, they will probably dry and shrink even more after being laid up in your wall, but this is a cosmetic problem which can be attended to a year or two down the road. Log-end expansion in the wall, while very rare, really spoils your whole day, even your whole month.

In between these extremes, Dave and Nancy Turkow near Rochester, New York, built a beautiful home of dry poplar (quaking aspen). Water collected on the round slab during construction and caused a ¼" crack to appear at the first course of log-ends, about half of the ½" chasm we experienced at Earthwood. But the slab dried out, rain hit the outside of the wall, the crack almost closed and the wall came back to plumb. The problem was never bad enough to necessitate the dismantling of any wall sections.

OTHER SOURCES OF WOOD

Cordwood masonry has been called "poor man's architecture," although there is nothing stopping a rich man from having a good time building a beautiful house by this method. Many writers in the field like to point out that the cordwood builder can use wood unsuitable for other types of construction, where long straight pieces are needed. Here are several potential sources of existing (and, in some cases, partially dry) cordwood:

Deadwood in the Forest

A large woodlot which has not been tended for several years may yield cords of dry deadwood in the form of fallen or leaning trees. If the bark is still on the wood, however, the chances are high that rot has already set in. Trees and dead branches without bark may well be dry and sound, however, as the first cut with a chainsaw will reveal. Admittedly, it would require a considerable woodlot and quite some labor time to supply enough cordwood by this method, but a couple of weeks of hard (but not unpleasant) work might make it possible to build that same season.

Warning! Standing dead trees, even "stringy" elms, are known as "widow makers" in the logging industry, and with good reason. Large limbs can break off during the felling process, due to vibration. Hardhats, of course, would be mandatory during the felling of such trees, but the only completely safe course is to avoid them altogether in favor of wood which is already on the ground.

Logging Slash

Another source of deadwood is to clear up the slash left after logging operations. Big loggers can't be bothered with short logs, tops, twisted pieces, and so on, and often leave them scattered on the lot. The debris that has managed to stay clear of the ground might make excellent cordwood. The owner of such a woodlot might be glad to have someone come in to tidy up the lot. Cordwood construction, then, can be beneficial to the environment. Jack Henstridge points out that by taking out the *waste* wood you give the other trees a better chance to grow. Jack also points out that fire-killed wood makes excellent log-ends; if you happen to be near an area that suffered a forest fire a year or two earlier, you may be spared considerable drying time.

Recycled Utility Poles

I have heard of at least three houses made with recycled utility poles that had rotted at the ground but still had 20' of good dry cedar or spruce just begging to be cut into cordwood. Cliff Shockey's two houses in Vanscoy, Saskatchewan, are made of old poles, and are superinsulated besides.

Reject this source of wood if it is saturated with creosote or other objectionable chemicals.

Split Rail Fences

The best material for log-ends that I've found is old split-rail cedar fences. Not only are they dry—they probably achieved ambient moisture content fifty years ago—but their random shapes make it easy to lay up a wall with consistent mortar joints. Finally, if these old cedar log-ends are left protruding from—or proud of—the mortar matrix, the rustic character of the old rails is accented. The effect is that of a very old wall, except that it is in perfect condition and will stay so for a very long time. Now, I realize that these old fence rails are a rather specialized item and are sought after as decorative fencing in the suburbs, but I was still able to buy them in northern New York in

1979 for $25 per 16" face cord, cut to length but not delivered.

The point is that log-ends are where you find them. Cliff Shockey found them on nearly treeless plains in the form of old utility poles. Even if you have to buy wood for the job, it may well be a bargain.

Wood Cutters

Firewood and pulpwood merchants, as well as suppliers of fence posts, are great potential sources of wood. In some cases, the wood is already at or near the right seasoned condition for building. Fence post makers often have short pieces left over from their operation—even cedar. They can be bought and trimmed to size. Even if you buy firewood or pulpwood at $100 per full cord, a "six cord house" could have its walls built for $600 plus the cost of the mortar and sawdust.

HOW MUCH CORDWOOD DO YOU NEED?

Estimating cordwood required is a simple matter and, as it is common to all three methods of construction, this seems as good a time as any to explain the calculations involved. The most convenient measurement of cordwood is . . . *cords!* Here, it is of paramount importance to differentiate between a *true cord*, which measures 4' high by 4' wide by 8' long, and a *face cord* (also called a *run*) which is 4' high by 8' long by whatever length of log the merchant is selling: foot-length, 16", 18", whatever. This can get very confusing, so we will stick to full or true cords in our calculations.

Incidentally, as a true cord has between three and four times as much wood as a run of "foot-wood," always make sure of a dealer's definition of "cord" before buying wood for any purpose. A true cord is not exactly four times more voluminous than a 12" face cord because the foot-length stuff packs much tighter than the more irregular 4' lengths. A full cord cut into 12" log-ends and restacked may yield only three to three-and-a-half face cords, depending on the straightness of the original logs (2-13).

This compression of the piles in the restacking can be incorporated into our calculations very nicely, as the loss of wood will be more than made up by

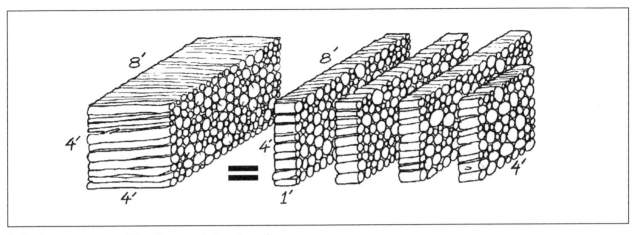

2-13.The strange case of the disappearing cordwood!

the gain of wall area through the mortar joints in a cordwood wall. So, if we deal in 4' cords—a convenient size for handling in the woods—we can disregard both the mortar area of the finished wall and the disappearance of wood when it is cut short and restacked. Nice. And there should be wood to spare, which is a whole lot better than being short of seasoned log-ends.

One good thing about cordwood masonry is that the leftover material makes darned good firewood.

The chart shown below can be used to determine the number of full cords of wood required to build an 8'-high wall with an inside perimeter of 120' at various wall widths. The wall widths given

have been chosen because log-ends at these lengths can be taken from 4' cordwood with no waste. Because ¼" is lost with each cut of the saw, the average "real" wall width is slightly less than the full widths given. Care should be taken that the wood is hauled out of the forest in lengths equally divisible by the length of log-end required. Otherwise, you run into problems of wastage and unnecessary sacuts. It is assumed that wood "loss" due to tighter stacking at the shorter length is equalled by wall area gain because of the mortar joint.

Note that the measurements given are *inside* measurements. On rectilinear structures, the corners will be either post-and-beam or some sort of built-up

Number of full cords of wood required to build an 8'-high wall with an inside perimeter of 120'

A	B	C	D	E	F	G
			Wall area			
	Gross		covered		Adjusted	Adjusted
	internal	Face	by full	Full	wall	full cords
Width	wall area	cords per	cord	cords	area	needed
of	(120x8)	fullcord	(Cx32)	needed	(.85B)*	(F/D) or
wall	(sq ft)	(48"/A)	(sq ft)	(B/D)	(sq ft)	(.85E)
8"	960	6	192	5	800	4.167
12"	960	4	128	7.5	800	6.25
16"	960	3	96	10	800	8.333
24"	960	2	64	15	800	12.5

*Assumes 15 percent of wall devoted to windows, doors and framing.

square timbers. Either way, they do not enter into the cordwood calculations. With a curved wall, the mortar joints are wider on the exterior than on the interior, but only fractionally. Wider mortar joints are okay, but narrow ones are unacceptable when cordwood is load-supporting. Also, the use of interior measurements means that we are calculating *actual usable internal square feet*. With the thick walls of cordwood construction, usable square footage is much more meaningful than gross square footage calculated from external dimensions.

The example given, 960 sq ft of gross internal wall area, is typical of a 30' internal square (900 sq ft of floor area) or a 38.2' internal diameter (1146 sq ft) round house as per Figure 1-5, assuming eight foot (8') high walls. You can make your own calculations in a similar manner, plugging your own gross internal wall area into the chart.

If you prefer to deal in face cords, simply divide the gross internal wall area by 32, then take 85 percent of the result (door and window adjustment) to arrive at the number of face cords needed. You will have enough to be quite fussy with your wood, rejecting twisted, misshapen or otherwise substandard log-ends.

INSULATION OPTIONS

The last of the components of a cordwood wall—insulation—is of equal importance to the mortar and the log-ends themselves, providing you want to stay warm in the winter and cool in the summer. Simply leaving a dead-air space between the mortar joints is not good enough for a home, although it may well be good enough for a barn.

We insulated our first three cordwood buildings with strips of fibreglass insulation. We don't do this anymore. The problems with the fibreglass were that it was expensive, not as sound environmentally, more time-consuming, and has the potential for "matting down" (and not springing back) if it gets wet. But the worst problem was that fibreglass was nasty to work with. We had to cut batts of insulation into narrow strips, using a sheetrock knife. Fine glass fibres could be seen floating in the sunshine, some-

times getting into the eyes—a real irritant—and probably into the lungs.

Thankfully, Jack Henstridge introduced me to a much better option—treated sawdust—which we have used ever since, and now recommend highly. The sawdust is cheap, makes use of a natural waste product, is easy and safe to use, and has an R-value equal to fibreglass. We insulated all the cordwood walls at Earthwood with one large dump-truck load of softwood sawdust from a local sawmill at a cost of $75. In addition, this load also provided the sawdust used in the mortar.

The sawdust is "treated" with two to three spadefuls of lime for each wheelbarrow of sawdust. Screening the sawdust is optional; it is a little easier to work with if you pass it through a half-inch screen. Work the lime into the sawdust with your spade or hoe until the mixture has turned a uniform white color. I find it convenient to keep a special wheelbarrow on site just for the sawdust insulation, making a new mix whenever I run out.

I keep my sawdust pile covered against rain. If the sawdust is just a little damp, that's okay. The lime will help to dry it. Similarly, fresh green "sappy" sawdust from the sawmill is okay. But don't let your pile get saturated. It is best if the sawdust is dry or only slightly damp.

The lime protects against insect infestation and fungal growth, but has another advantage, as we have found. If the wall gets wet for any reason, or if the sawdust is somewhat damp in the first place, the lime will cause the mix to "set up" somewhat, making an insulation akin to beadboard when it dries out again. If the sawdust stays dry during construction, it has the characteristics of loose fill insulation such as vermiculite.

Incidentally, vermiculite and similar loose fill insulations such as Perlite® can be—and have been—used successfully. They will cost more money, obviously. I would avoid cellulite or blown-in insulations. Also, chunks of rigid foam insulation are difficult to work with and do not completely fill the void. Shredded beadboard has been tried without success; the slightest breeze blows the particles onto the wet mortar. If left in place, this will spoil what little bond there is between wood and mortar. Scraping it off all the time really puts a crimp in production.

3

Cordwood Masonry Within a Post-and-Beam Framework

BALLOON FRAMING

This method uses cordwood as "nogging" or infilling within a framed wall. Although most frequently used within a heavy post and beam frame, wide balloon framing can employed, such as with two-by-ten or two-by-twelve studs, four feet on center (4' o.c.). Many of the old cordwood buildings in Wisconsin were built this way. A corner option would be to make corner posts from dimensional timbers, as shown in 3-1.

3-1. A box post.

SINGLE-WIDE POST AND BEAM

Most houses using cordwood infilling make use of a regular post and beam framework. A great advantage of the style is that the whole frame can be completed, including the roof itself, before any cordwood work is done. We did this ourselves at Log End Cottage.

At Log End Cottage, the largest span between posts is 8', measured on center. This situation occurs at both gable ends, so we chose our largest and best pieces as corner and gable support posts. Along the side walls, no span is greater than 6' on center. Our original intention was to add a sod roof in the future and we wanted to be sure that the framework would support the tremendous weight. The roof finally ended up with cedar shingles, so we really didn't need to build the frame anywhere near as strong as we did.

Besides corners, posts are needed to frame doorways and large windows.

Old barn timbers, if available, are ideal companions for log-ends. They are dry and, provided care is exercised in their selection, incredibly strong. Avoid punky or insect-infested beams. Many old timbers have had woodworm in the past, but the worms have long since abandoned the wood after it became too dry for them. No need to fear these timbers, but make doubly sure the worms are gone by excavating a little with a knife. Judging the structural quality of old beams is an acquired skill, but play it safe while acquiring it: if in doubt about a beam, don't use it.

Old barn beams are becoming increasingly hard

3-2. (above) Post-and-beam framework, south gable end.

as some have suggested, in an attempt to fasten the posts to the cordwood mortar. When the newly milled posts shrink, they will rip the mortar with those nails imbedded in it. Caulking will be more difficult and will show more.

Figure 3-2 shows a view of the framework. After the photograph was taken, we added diagonals along the side walls just as we did at the gable ends. The diagonals are paired three-by-tens with a 3" insulated space in between. They are, therefore, exposed both inside and outside. Diagonals add rigidity to a post-and-beam structure because they form triangles. In modern framing, plywood serves the same purpose. My thinking now is that the diagonals, even though they are very attractive, are more trouble than they are worth. The cordwood masonry will supply all the rigidity required. However, temporary diagonals should be employed until at least one panel of cordwood infilling is completed on each wall.

As a point of interest, our floor joists and roof rafters are also old three-by-tens, 24" on center. We were lucky to find excellent buys on three-by-tens from people who were demolishing an old school and an old hotel not far from our homestead. *See 3-5 to 3-7.*

All posts should be checked or toenailed into a wooden plate fastened to the masonry foundation. It is imperative that the top ends of all the posts are tied together by the use of another plate. In Log End

to find. It is likely that you will have to go with new rough-cut material made at a local sawmill. The advantage is that you can give the sawyer an exact materials list for your house, and know that you'll have good sound members for each part. The disadvantage is that the posts and beams will shrink away from your cordwood panels after a couple of years. The best way to handle this is to run a bead of clear silicone or silicone-based caulking along the timber a couple of years after the house is finished. (Do it too early and you'll have to do it again, but that's not so bad; it is easy to do.) I do not put nails in the posts,

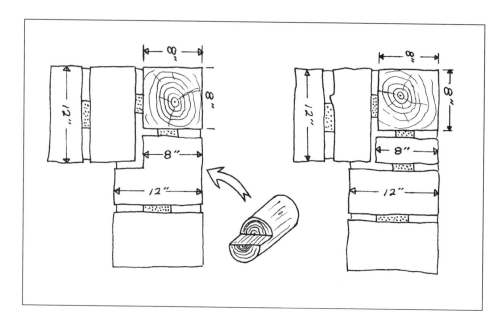

3-3. Two methods of using an eight-by-eight post in a 12" wide cordwood wall. In the second method, a special log-end is cut, split and placed as shown. These methods can be combined in the same corner according to the wishes of the builder.

Cottage we used heavy barn timbers, as much for aesthetics as anything else. Doubled two-by-eights or two-by-tens would certainly accomplish the same thing. With the tops of the posts firmly tied together by a plate, the pressure exerted by the masonry panels is resisted.

At Log End Cottage we used two large barn beams to tie the long east wall to the west wall. These beams fly from plate to plate through the kitchen dining area—which has a cathedral ceiling—and they make a convenient place to hang pots, pans, mugs, and the like. Because the eight-by-ten ridge beam is firmly supported by four posts, and because the roof rafters are notched into the ridge beam as well as birdsmouthed upon the wall plate, the two tie-beams may not be strictly necessary. However, the wall plates are pieced together (two pieces on the east side, three on the west), and I am glad I installed the tie-beams for the extra rigidity.

Heavy framing with old barn timbers is just as physically demanding as building a traditional log cabin, so we measured and cut each section on the ground and then reassembled it in place with the help of friends—but there were only nine beams in our cottage that required help, compared with 60 or more long logs in a log cabin of comparable size.

Another drawback to the use of old timbers is that every one is of a slightly different dimension. In fact, the size will vary from one end of a beam to the other by as much as 1". Obviously, a lot of extra measuring, figuring, and notching is necessary. This is tedious work, but worth it as it's good to be square and level when the time comes for putting up the roof rafters.

It may be desirable to build the roof before any cordwood masonry is laid up so that you'll be able to work in wet weather, and you give the cordwood extra drying time. The post-and-beam method is the only one of the three which allows this option, but if it is taken, temporary diagonals will need to be nailed up on the outside frame for rigidity until the cordwood is laid up. At our Cottage, the roof was completed before a single log-end was placed in the framework.

Ordinary framing, roofing, and roof-insulating techniques may be employed once the top plate is secure. It is beyond the scope of this book to enter into all phases of building. There are many good manuals on timber framing, plumbing, electric wiring, et cetera. I will not embark on these subjects,

except where the cordwood masonry technique demands special considerations.

Our Log End Cottage had walls 9" thick. I now consider this inadequate in an area with more than 3000 heating degree days. (This corresponds, roughly, to all the lands north of a line connecting San Francisco, Las Vegas, El Paso, Memphis, Atlanta, and Raleigh.)

THE DOUBLE-WIDE FRAME

Jeff Corra of Parkersburg, West Virginia (about 5000 degree days), solved the narrow wall problem by building his post-and-beam frame using doubled eight-by-eight timbers. His red-and-white oak log-ends are 16" long.

At the corners, two eight-by-eight posts stand side by side, not four, as might, at first, seem to be necessary. *See* Figure 3-4.

Jeff used a lot of heavy timbers, but his house will grace the city of Parkersburg for many long years into the future. For some, the double-wide approach to framing may be a daunting prospect, indeed. For those building in an area where a 12" wall might be sufficient (less than 5000 degree days), I offer two different methods of using a single eight-by-eight as a corner post (3-3). Two doubled four-by-six timbers are plently strong enough as intermediate posts between the corners, used to frame doors and windows, perhaps, or simply every six or eight feet around the house. The reader might work out other material-saving methods of using heavy timbers in a thicker wall. Remember: whatever type of frame is chosen, it should be designed so

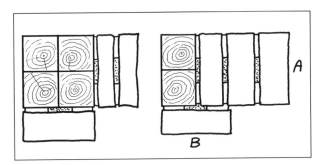

3-4. Only two (not four) eight-by-eight posts need stand at the corner, if Wall A is built before Wall B.

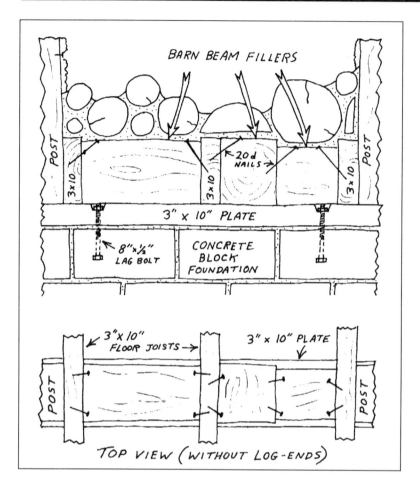

BARN BEAM FILLERS

POST

3×10

20d NAILS

3×10

POST

3×10

3" × 10" PLATE

8"×½" LAG BOLT

CONCRETE BLOCK FOUNDATION

3"×10" FLOOR JOISTS

3" × 10" PLATE

POST

POST

TOP VIEW (WITHOUT LOG-ENDS)

3-5. In Log End Cottage, the author used three-by-ten floor joists, 24 inches on center. The joists cantilever out over the foundation plates (also three-by-tens), supporting the front porch and the back firewood deck. It is convenient to fill in the space between joists with segments of barn beams left over from the framework. The barn beam fillers add support to the joists and establish a fairly flat surface upon which to be laying log-ends. The barn beam sections are hidden from view by decking and inside flooring, except in the basement. The three-by-ten plate is bolted to the foundation by 8' x ½" lag bolts which are well cemented into the last course of concrete blocks. The author used three bolts for each ten-foot section of the plate, leaving the threaded end 3 inches above the top of the block foundation.

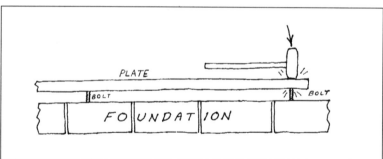

PLATE

BOLT

BOLT

FOUNDATION

3-6. Find out where to drill for the bolt holes by laying the plate upon the bolts and whacking the plank stiffly with a hammer. Then remove the plate and drill the half-inch holes where the bolts have left depressions in the wood.

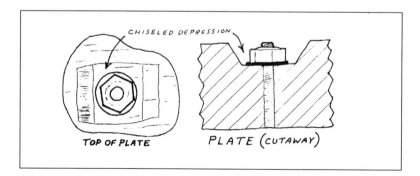

CHISELED DEPRESSION

TOP OF PLATE

PLATE (CUTAWAY)

3-7. It will be necessary to chisel enough wood out of the top of the plate to receive the washer, nut, and socket attachment.

that it can do all the load-bearing independent of the cordwood nogging. Therefore, the tops of all the wall posts, corner and sidewall posts must be tied together with a plate beam system capable of supporting the roof. If in doubt about your particular design, consult with an architect or engineer.

Before building the framework, become familiar with the structural considerations of timber framing. There are many books on the subject, but one I particularly like—well enough to offer it at the Earthwood School—is *Timber Frame Construction; All About Post-and-Beam Building* by Jack Sobon and Roger Schroeder (Garden Way, 1990).

I must admit that in the post and beam structures I've built, I have not taken the time nor developed the skills to make all those beautiful mortise-and-tenon joints used in the old-time barns, and beautifully illustrated in the books. Some of our students have, though, such as Geoff Huggins of Winchester, Virginia. See his work in Chapter 6. No, I'm afraid that I've "cheated" somewhat, by using truss plates, toe-nails and long log cabin spikes to hold my frames together. I try to hide

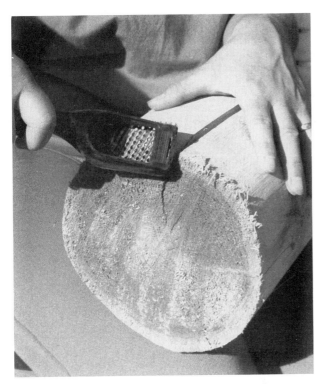

3-8. Jaki cleans the edges of a log-end with a Stanley Surform Shaver.

these things as well as possible, and most people are impressed enough with the heavy timbers themselves so that they don't notice anything amiss. Purists—*craftsmen*—are entitled to look down their noses, however.

FINAL PREPARATION OF LOG-ENDS

Although already barked, split, and delivered to the work site, the log-ends may need some final attention before they are laid up. If northern white cedar—my favorite—is chosen, then it will be necessary to smooth away the hairy edges at the ends of the logs, an inevitable result of the cutting process. These fibres get in the way of pointing and inhibit the formation of a good bond. The best tool we've found to do this is a Stanley Surform Shaver (tool number 21-115).

Cylindrical log-ends, especially those over 4" in diameter, should have fairly substantial checks in them from drying. Jaki and I like to stop drafts -rom passing through these primary checks after the wall is up. We have found the best material to stuff them is ordinary newspaper strips soaked in a bucket of water. We stuff this paper into the checks with a flat-head screwdriver, about an inch or two in from each end of the log-end, so that the stuffing is not readily visible. The natural check still shows, but drafts are stopped. This process can be done after the wall is built—and *should* be if further radial shrinkage is expected—but it is easier to do a good job before they are laid up. The clear caulking solution already mentioned is an option after the walls have seasoned a year or two.

Finally, it is possible that the ends of your log-ends have greyed from weathering in the woodpiles or from the bacterial action on the wood sugars, especially prevalent with poplar. If you have a circular saw available, you may wish to clean one end by cutting off the thinnest possible slice, just ¼" or so. (Wear eye protection.) Lay up all the greyed ends to the outside and the fresh clean ends to the inside.

Another approach to this problem is to brush one end of each log-end with a 30 percent bleach solution the day before laying it up. This cleans the end and kills the bacteria. The blackened end will not return. The only drawback is that the bleach washes out some of the natural color of the wood at the same time. At Earthwood, we have employed a third solution. After the wall was cured, we sanded the interior surface of each log-end with a high-speed 5"-diameter disk sander. This looks great. Be sure to wear protective breathing and eye equipment.

START WITH A CLEAN FOUNDATION

With the post-and-beam style, it is quite likely that your first course of masonry will be on a wooden plate or plate beam, as has been already shown. If this plate will not be part of the interior space, it could and should be made of pressure-treated lumber. This would be the case if there is a crawl space only under a wooden first floor. Another possibility is that you might anchor all of your posts to a solid concrete foundation. This is done in the same way as the heavy door frames are fastened to the floating slab in a round house, described in the next chapter.

In any case, it is necessary to start with a clean dust-free foundation to maximize the bond with the mortar. With a concrete foundation, I find it sufficient to dampen and brush the area to be worked upon just a few minutes before starting work. If a wooden plate is used, the bond can be maximized by the application of a coat of Acryl-60 or equivalent bonding agent an hour before working on a particular section. The Acryl-60 approach can be taken with a concrete foundation, too, but isn't really necessary unless the surface is glassy-smooth.

BUILDING WITH LOG-ENDS

Mortar mixing proportions and how to mix the mortar have already been discussed in Chapter 2. Now we will assume that you have delivered the wheelbarrow of mud to the site and are ready to lay up the wall, or, in this case, fill in the post-and-beam framework with cordwood nogging.

Okay, the foundation is clean, the mud is here, and a good variety of log-end sizes is within reach. Let's go.

With your rubber gloves, reach in and grab an American Standard Handful of mud. Place it down on your foundation and shape it into a bed of mud one inch (1") thick. You will learn to shape the bed of mortar with your thumbs as shown in 3-9. The width of the mortar joint is as per 3-10.

The procedure is to lay out a double bed of mortar as per 3-9, 3-10, and 3-11. We used to lay mud with a trowel but now place the mortar faster and more accurately with rubber gloves. If you are already a mason used to working with a trowel and wish to continue that way, fine. But if you are a mason, you will know that gloves must be worn, even while using a trowel.

Cordwood masonry is time-consuming, as you will find out. But there are efficiencies which you can learn which will help to speed up the process. One of the most important is to develop efficiency in the handling of the three primary materials: mortar, insulation, wood (in that order.) Avoid constant unnecessary

3-9. Thumbs and fingers work together to put down a bed of mortar.

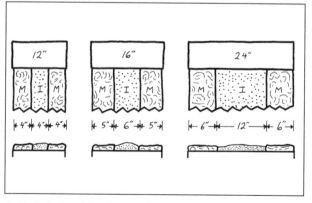

3-10. Relationship of mortar bed width to insulation for walls 12", 16", and 24" thick.

switching back and forth from one material to another. On the first course, for example, lay down six feet of the double bed of mortar, perhaps the entire width of a panel between posts. Then, do the insulation for the whole six feet, then the log-ends. Don't get caught up building a little mountain peak area of wall thinking that the wall is somehow going to go up faster that way. Build in courses, both for efficiency and for sound building practice. Now, as the weather man says, for the details.

Mud

Consistency of thickness and width are the important considerations here, as per 3-9 and 3-10 above. At the school, we have students use a "cheat stick," shown in 3-13. The stick is just a scrap piece of smooth straight wood with a graphic in-scale crayon drawing showing the location of mortar and insulation over the width of the wall. When students start skimping on mortar, we place the cheat stick right on their mud before they get a chance to cover the work with a log-end. Instantly, they see the error of their ways. Some students like to use the cheat stick for the first hour or two, until they develop a feel for the proportion. My cheat stick is made from a one-inch (1") thick scrap, so it can even be used as a gauge for the thickness of the bed.

Incidentally, 99 percent of the time the error in proportion is to skimp on the mortar, usually on width. I expect that there is some deep-rooted sense of economy behind this, but, please, don't do it. Better to mix an extra 20 wheelbarrow loads of mud at your house, but have a strong, pleasing, easy to point wall.

Insulation

Next, using a small bucket—it is worth finding a one or two gallon plastic paint pail with a little pouring lip on it—pour the insulation in between your mortar beds, using long sweeping motions. On a round house, be sure to work from the inside, so that the sweep of your bucket corresponds to the curvature of the wall. Do not straddle the work while you pour; this will cause the insulation to spill onto the mud, necessitating time-wasting cleaning operations. The right amount of insulation will come with experience. If you put in too much, it will actually inhibit the proper placement of the wood; the log-ends will

3-11. The mud in place, first course.

3-12. Pour the insulation in to the central cavity with a sweeping motion. Use a bucket with a pouring spout. With a curved wall, it is easier and tidier to work from the inside of the curve.

3-13. The "cheat stick" will help you along until the proportions become second nature.

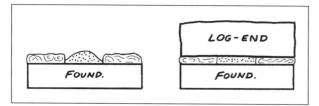

3-14. Cross section of insulation placement.

3-15. The wood is set in place. Leave room for pointing between log-ends. Use a variety of sizes to create "hills" and "valleys." *See also* 3-16.

rock on the insulation. If too little is used,there will be voids in the wall, not too bad if they aren't too frequent. Obviously, the R-value is somewhat lessened. My own method is to pour the insulation in so that it takes on the cross section shown in 3-14. By this method, the log-end will press down on the crest of the sawdust bead, thus filling the voids.

Log-ends

The last material is the wood. Lay the log-ends firmly in place, tying the inner to the outer mortar joints. Use a slight vibrating motion back and forth to create a suction bond with the mud. This helps prevent accidentally knocking the log-end loose while working and helps maximize the final friction bond between the wood and mortar. (One or two overly cautious builders have taken the trouble to hammer old nails around the log-ends where they come in contact with the mortar, a total waste of time, in my opinion.)

Leave a space between log-ends a little greater

than the width of your pointing knife; about an inch is good. Make sure you place log-ends over the entire section of mortar which you have laid down. It is all too easy to get caught up with some detail at one end of the section and neglect the mud which you'll have put down a half hour ago. The mud stiffens and it becomes a lot harder to place the log-ends properly.

It is very important, when laying up the first course, to get away from the flat plate and into a random pattern as quickly as possible. If the first course is laid with log-ends of the same diameter, there is a danger of getting stuck in a pattern which is hard to break and, paradoxically, hard to maintain. Esthetically, masonry looks good if it is totally random or if it is very carefully laid up in a pattern. It looks bad if someone tried to incorporate a pattern and failed; and it looks bad if a seemingly random wall becomes patterned.

I like the random look: mistakes are less obvious, the wall is strong, and it incorporates all different diameters and shapes of log-ends. This latter feature helps ensure that the various sizes of log-ends in the pile are depleted at an equal rate, leaving a good selection right to the end. One of the little tricks of stone masonry, which applies equally to cordwood, is to keep a variety of pieces handy. Probability dictates that the right log-end is almost certain to be in a pile of, say, 50 random pieces, but sometimes, especially near the framework, it will be necessary to split a specific piece from a clear-grained log-end.

Good pieces to start with are log-ends split in half, or, for corners, split in quarters. Another handy shape is what I call a "slat-end." A slat is the first piece taken off a log when it is being squared for lumber. We acquired a couple of pick-up truck loads of slats at no cost when a friend was having cedar milled.

Voilà! The first course is done. Succeeding courses follow the same pattern: handle mortar, then insulation, then wood. The only real difference is that the mortar beds are no longer flat, but take on the contours—the "hills" and "valleys" formed by the log-ends of the first course. If there is a particularly high spot in the course, such as might occur when a large round log-end is laid, I do not cover the top of it with mortar right away, as this mud will probably not be covered with wood in the next course. The mud gets stiff and has to be removed. After the first course, always think in terms of filling in valleys, not climbing over hills. By this method, the old valleys become new hills, the old hills become new valleys. This is part of

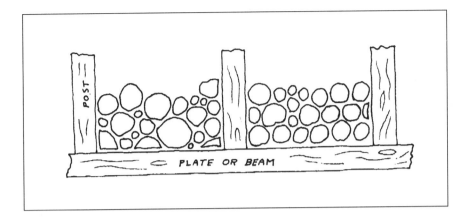

3-16. The first course establishes the random pattern.

3-17. (left) Next, we "fill teeth" with mortar. This is our term for stuffing some mud between adjacent log-ends, which protects them from being moved accidentally. Also, this is the easiest opportunity to place mud in that location.
3-18. (right) Proceed with mortar, following the hills and valleys of the first course...

3-19. ...insulation...

3-20. ...and more wood.

what I mean when I say that the wall "builds itself."

Drawing 3-16 shows the importance of the first course. Once a random line is established, there will no longer be true courses. Rather, the mason will be dealing with individual spaces of varying sizes. The wall will almost build itself, calling out for the next piece required. This is as it should be for a truly random wall. Also, it helps to keep mortar joints small—¼" to 1" thick is good. Any smaller makes for difficult pointing; any larger and it's hard to insulate, doesn't look good and wastes mortar. Of necessity, some joints will be 2" thick or more. Not to worry.

When I suggest that the wall will build itself, this is not meant to imply that the builder can turn his brain off for the rest of the panel. Sometimes chance will throw the wall into an unwanted pattern. The first warning is when five or six log-ends of the same size appear in the same location. Log-ends of the same size fit beautifully into a hexagonal pattern, like a honeycomb, but this pattern will soon deplete one size rapidly.

Don't waste time looking for log-ends before the mud is down. Until you put the mortar joint down, you can't readily see the size and shape of the next log-end you need. Mud first, then wood. This is the other part of what I mean when I say that the wall builds itself.

The exception to the commentary above is when you have available thousands of round log-ends of the same diameter. In this case, you will build in straight courses, lapping the previous course as shown in 4-1 in the next chapter. Beautiful homes have been built this way by Cliff Shockey of Vanscoy, Saskatchewan, and Ken Campbell of Idaho Springs, Colorado.

Another exception to the "keep it random" rule would be if the builder deliberately wanted to feature some planned configuration in the midst of random background. This type of planned deviation can be extremely effective and is discussed in Chapter 7.

Another good reason for the mason to stay awake is that chance will rarely provide good opportunities to lay the 9" to 12" diameter log-ends. Large ends require a "cradle," and this requirement will have to be provided for once in a while. This won't slow down construction, however, as one 10" log-end gets a lot of wall up in a hurry.

Other points to remember during the laying-up:

• The builder should lean over the work once in a while, eyeballing it up and down, right and left. *Is the work plumb?* Large panels should be checked with a level now and again.

• The builder should stand back from the wall once in a while and ask himself, "Is the work balanced? Does it please the eye? Why? Why not?"

• *One mortar joint should not be placed directly over another.* This basic rule of masonry is even more important when building with log-ends because there is little bond between wood and mortar.

• The going gets tough near the top plate. It is a

LAYING UP LOG-ENDS: A STEP-BY-STEP GUIDE

3-21. (left) This sequence makes use of a demonstration panel constructed to simulate a real panel in a post-and-beam house. *See also* page H of the color section. 3-22. (right) Lay a double bed of mortar, most easily done with rubber gloves.

LAYING UP LOG-ENDS: A STEP-BY-STEP GUIDE *(Continued)*

3-23. (left) Insulate with treated sawdust or vermiculite. 3-24.(right) Place the log-ends with a slight vibrating motion to get a good suction bond. Use random sizes.

3-25. (left) The mortar beds of the second course follow the hills and valleys created by the log-ends of the first course. Note the large valley or "cradle," soon to be useful in setting a large log-end. 3-26. (right) Insulate the second course.

3-27. (left) A small container, such as a tin can, helps in tight spaces. 3-28. (right) The cradle formed on the first course gives an opportunity to set this 12" diameter log-end. It helps to plan for the larger pieces.

3-29. (left) This log-end is made 6" longer than the others, and will be used as a shelf. The exposed part is sanded.
3-30. (right) The going gets tougher near the top of a post-and-beam panel. This log-end is too big, but it can be marked with a pencil and ...

3-31. (left) ...split to the right size and shape. Choose clear-grained logs for ease of splitting, and don't forget eye protection. 3-32. (above) The split corner piece went in nicely, and all the log-ends are in place on the top course. Because of poor access, it is no longer possible to place loose fill insulation in the panel. This is the one place in the wall where we still use fibreglass insulation, forced into the topmost cavity with a pointing knife.

3-33. (left) To fill the final gap, push mortar off the back of a trowel with a pointing knife. The fibreglass insulation already stuffed in will act as a backing. 3-34. (right) The panel is pointed with a kitchen knife which has had its blade turned up slightly. Pointing both strengthens and beautifies the wall.

3-35. (left) The finished shelf. 3-36. (right) The finished demonstration panel. Lord Louis Logend likes his new home.

great help for someone to help the mason from the other side of the wall, by stuffing insulation, laying bedding, and guiding pieces into place. This is one area where I still use fibreglass insulation instead of sawdust, as shown in 3-32. Sometimes, in close quarters, a log-end will only fit in from the other side. A good way to fill the final joint is to push the mortar off the back of the trowel and into place with the pointing knife.

• The builder should not be afraid to use odd-shaped pieces. They'll add interest to the wall.

POINTING

This is one of the most important parts of the whole job. Many people think that pointing is purely decorative, to smoothen rough cement. Not so. Proper pointing greatly *strengthens* the wall by tightening the joints and reducing the chance of mortar cracks. The pressure of the mortar against the wood under the pointing knife gives the best chance for a good bond. Therefore, a fair amount of pressure should be applied to the knife as it is drawn along between log-ends, but not so much as to push it into the insulation cavity.

But, yes, pointing is *decorative* as well and is important to the finished appearance. There are many kinds of pointing and tools with which to point but Jaki and I feel that the wood, and not the mortar, is the predominant feature of cordwood masonry, so we accent the wood with a simple recessed pointing, the log-ends ¼" to ¾" proud of the

mortar. Recessed pointing helps provide the interesting relief of a log-end wall. When using cylindrical log-ends, there is another argument against fancy pointing, such as araised V-joint: the joint is very small where the log-ends are tangent and wide in the space between three ends, so a time-consuming job becomes tedious and difficult.

Pointing is done while the mortar is still plastic. On hot days care should be taken that the pointing is not put off too long. Usually, there will be enough mortar squeezing out of the joints with which to point, but it is a good idea to save a little at the end of a day's laying in case there are substantial holes to fill. A really big hole should be stuffed with insulation first. Pointing is shown in 3-34.

We've made many a good pointing tool from old kitchen knives by bending the last ¾" of the blades to a 30° angle from flat.

BUILDING A POST-AND-BEAM SAUNA

To conclude this chapter, I offer complete plans for a very simple post-and-beam structure, plans which we used to build Log End Sauna, shown in 3-37 and 3-38. It is designed as an outdoor sauna, but the plan could easily be adapted to a wilderness cabin or garden shed, or might even provide temporary shelter while a larger structure is built, to be used as a sauna or shed later on. Swedish owner/builders have for years adopted the strategy of building their sauna first (an important priority to the Scandinavian people), and moving into the building until their house is complete.

3-37. (above) A cordwood sauna. 3-38. (left) The sauna has a post-and-beam frame and a freestanding earth roof.

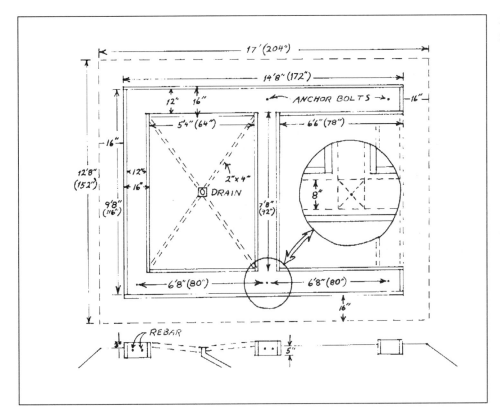

3-39. Shuttering diagram for the footings.

First, clear the topsoil from an area measuring about 12' 8" by 17'. Save it for use later on the roof.

Then, 8 cubic yards of sand is spread over a 12'8" by 17' area to a depth of 1'. The two volumes are equivalent. The *pad* thus created should be tamped while it is moist, to assure maximum compaction. The pad's purpose is to draw water away from the sauna, which is *floated* on the pad. This building technique, known as the *floating slab*, is one of the cheapest, easiest, and most effective means of building a frost-protected foundation for a cordwood structure. For a home, of course, the pad would be very much thicker, at least 18", as will be seen later.

Footing Forms

A track with the dimensions shown in 3-39 is dug in the pad with a flat shovel, and two-by-eight forms are set in place as shown. Two pieces of No. 4 rebar (½" stock) are placed in the track, to be drawn up into the center of the footing as the concrete is poured. If care is taken with the forms, and they are cleaned after use, they can be used as plates or for any of several other purposes later in the construction. Anchor bolts

should be set in each corner, where the posts will be. This eliminates the need for a floor plate.

As shown in 3-39, a 4" drain should be set into place in the pad before the forms are set in place. The forms should be oiled for easy removal and cleated with movable cleats as shown in 3-40. This type of cleat guards against bulging between the cor-

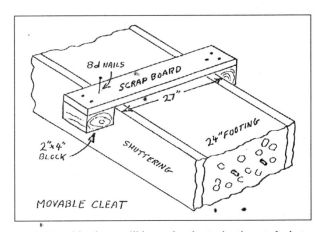

3-40. Movable cleats will keep the shuttering intact during the pour.

ners. That they are movable makes for easy screed-ing of the concrete.

Concrete

There are only 1.1 cubic yards of concrete required for the footings, so the con-crete could be mixed in a wheelbarrow, although a mixer would make the job easier. The mix should be 1 part Portland cement, 2 of sand, and 3 parts crushed stone (¾" diameter). Anchor bolts and rebar should not be neglected. After a day, the forms are removed, cleaned, and stacked aside.

The Floor

The floor should slope from all corners to the 4" drain in the center. This is not diffi-cult to do. Four two-by-fours, each measuring 4' 6", are set into the sand to a depth of l" (see 3-39). They should be set so that they start at footing level in the corners and slope down-ward about ½" to the center. This slope can be checked with an ordinary level.

The floor, 3" thick, can now be poured in four trian-gular sections. Each section can be screeded using the two-by-fours as guides, so that water will be carried to the drain at the center. Wire mesh reinforcing is advis-able, but not absolutely im-perative.One-inch Styrofoam® under the concrete floor is another option, one which will assure a warmer floor. If Styrofoam® is not used, it

would be a good idea to use sawdust concrete for the same purpose. It is made by replacing the crushed stone in the previous mix with an equal volume of sawdust, the kind that comes from a sawmill. The ingredients should be dry-mixed and then the water should be added. As with sawdust

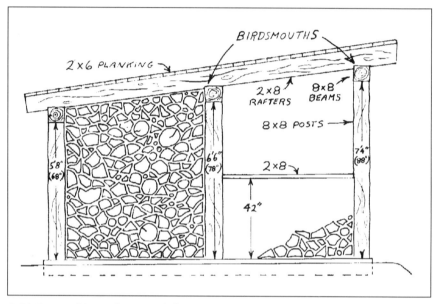

3-41. Post-and-beam framework for a cordwood sauna, side view.

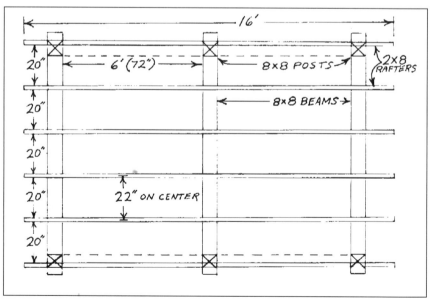

3-42. Roof support system for the cordwood sauna.

mortar, the sawdust admixture should be screened and soaked overnight before use.

Framework

Next, the framework can be built. Six eight-by-eight barn timber posts are stood up at the corners, with heights as shown in 3-41. Holes drilled in the bottom of the posts receive the anchor bolts. Three similar 10' barn beams span the three sets of posts, allowing a 4" overhang by the beams as an esthetic consideration. Temporary diagonal braces should be nailed to the framework for rigidity until the roof and cordwood masonry are complete. Six 16' two-by-eights make up the rafter system (3-42). They are birdsmouthed onto the support beams as shown in 3-41.

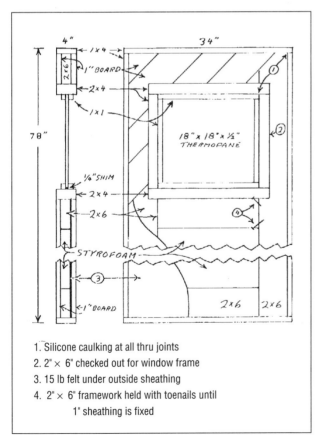

1. Silicone caulking at all thru joints
2. 2" × 6" checked out for window frame
3. 15 lb felt under outside sheathing
4. 2" × 6" framework held with toenails until
     1" sheathing is fixed

3-43. A good door for a sauna.

Roof

At our sauna, we made use of some planed 10' long two-by-six boards which we had on hand. With the 4" overhang on each side of the rafters, this worked out well with the rather strange spacing of 22" on center shown in 3-42. I recommend two-by-six tongue-and-groove planking as we have since used at the house and office at Earthwood. If ¾" plywood is used, another legitimate option, the plans should be adjusted to make use of 24" centers, which works out more efficiently with the 4' by 8' plywood dimensions.

We used an earth roof at Log End Sauna, and it has passed the test of time, over 12 years so far without deterioration. The earth roof construction is covered in Part Two: Earthwood. Another roof option for a shallow-pitched roof like this, though not so pleasing esthetically or environmentally, would be a surface of white mineral-faced half-lap roll roofing, applied according to the manufacturer's instructions. I would advise a minimum of 1" Styrofoam® over the subroof for insulation, covered with a layer of ½" plywood before applying the half-lap.

Flashing is applied next. The edge of the roof can be flashed with 50' of 10"- or 12"-wide aluminum flashing, folded down the middle to a 90° angle, so that half hangs over the edge to act as a "drip edge," and half rests on the roof. The flashing or drip edge can be nailed or stapled to the top roof surface, as it will be covered with a waterproofing membrane or half-lap roll roofing.

Waterproofing the Roof

For a discussion of applying the waterproofing membrane, see Chapter 16. The earth roof itself is explained in Chapter 19.

The roof structure described in this section will support a 6" sod roof with a 4' snowload. The two choices are to plant the roof in place, or to haul cut sods up and kick them into place with the side of the boot. The easiest way to plant the roof is to build the soil up to the required depth and spread the chaff from the hay over the top. Then spread a ½" layer of hay over the chaff and water the whole thing down. The hay will act as a mulch through which the new grass can easily grow. If sods are to be brought up, enough topsoil should be spread so that the addition of the sod will bring the depth up to 6". Keep the roof watered until well established.

Windows

Windows should be thermopane (double-glazed) and framed by two-by-eights. More on windows later in the book.

The Door

The door should be homemade and well insulated. A store-bought, finished door would really detract from the rustic appearance of this structure. People who make their own doors like to do it their own way, but I do include a cutaway view of the door which we used at Log End Cave as an example (3-43). This would make an excellent sauna door.

Let's see now, what have we forgotten? Of course, the walls! They are of cordwood, of course, as described in 3-21 through 3-36. The only special consideration applicable to a sauna would be to have one log-end, 4" in diameter, near the base of the back wall which can be removed to supply air to the woodstove. Similarly, a couple of larger ends could be removable at eye level, for ventilation or to peek out the side walls.

We have used two methods of installing removable log-ends for use with saunas. One is to wrap a cylindrical log-end with corrugated cardboard. Hold the cardboard on with elastic bands. After the wall is completely set, remove the log-end, then pull away the corrugated cardboard from the mortar. This is the only place in a cordwood wall that we ever allow the mortar to be continuous through the wall, so that insulation doesn't fall into the gap when the log-end vent is removed. Our other method of venting, used at the Earthwood sauna, is to set a 6" or 8" diameter by 8" deep ceramic thimble into the cordwood wall at the proper height. A cylindrical log-end of just under 8" diameter can be removed with a handle. A little rubber foam weatherstripping completes the seal in the closed position.

Benches should be made of nonresinous wood

3-44. Cylindrical log-end.

of low conductivity. Poplar is perfect. Hardwoods tend to burn the baby's bottom. The benches can be made of four two-by-sixes cleated together from below. Nails should not be exposed. A good idea is to make the benches adjustable so that they can be set at either the 24" or the 42" level, or can even be set at a gentle slope for reclining.

The sauna is heated with a woodstove. Yes, an electric heater made especially for saunas could be used, but a lot of the romance would be lost. Care should be taken that safe clearance is maintained from combustible walls (*both* stove and stovepipe). If this puts the stove too far out into the clear space, insulated wallboard made for the purpose can be used behind the stove. There should be a 1" air space between the wallboard and the cordwood wall.

The antechamber, which should face south, is used as a dressing room and relaxation area between sessions in the sauna. The floor for this area could be 3"-thick slabs of wood—very short log-ends, really—set into the sand. If both sides are sealed with two or three coats of polyurethane, they will last a long, long time. Short cordwood walls act as a windbreak and help this area serve as a sun trap. If desired, the antechamber could be screened in for summertime use and eventually closed in for year-round use. If this is anticipated, it would certainly be worth going through the little extra work required to complete the footing between the two posts at the south end of the structure, as we did. Removable Plexiglas windows could be used to cover over the screened areas. A sliding glass door could be incorporated between the two southernmost posts. But we're getting kind of fancy. If kept simple, the sauna can be built quickly and for very little money.

4

Curved Walls

WALL THICKNESS

As far as laying up the individual log-ends is concerned, there is very little difference between the curved-wall method and rectilinear methods: the mortar mix is the same and the wall thickness recommendations are the same. About the only difference is that the mortar joints will be wider on the outside of the curved wall than on the inside.

Of course, this difference is greater on thicker walls than on thinner walls. Similarly, the difference is greater on small-diameter houses than on larger. Here are six examples which illustrate these relationships.

House	Wall Thickness	Outside Diameter	Inside Diameter	Difference in Mortar Joints
A	12"	32'	30'	⅜"
B	16"	32' 8"	30'	½"
C	24"	34'	30'	¾"-1"
D	12"	40'	38'	¼"
E	16"	40' 8"	38'	⅜"
F	24"	42'	38'	⅝"

With sheds, saunas, and other small-diameter buildings such as "Mushwood," illustrated in Chapter 6, this difference in size of inner and outer mortar joints is even more pronounced, as can be seen here:

Building	Wall Thickness	Outside Diameter	Inside Diameter	Difference in Mortar Joints
Sauna A	8"	10' 4"	9'	1⅛"
Sauna B	12"	11'	9'	1⅜"
Shed A	8"	15' 4"	14'	⅝"
Shed B	12"	16'	14'	⅞"
Mushwood	12"	22'	20'	1"

It is important that the inner joint be thick enough to aid in load-supporting, so the wall should be built with the log-end spacing established on the *inside*. The outer joints will be thicker as shown in the chart. This chart assumes a log-end with an average diameter of 5". If smaller ends are used, the difference in size between inner and outer joints will be smaller, and if bigger ends are used, there will be a greater difference.

The difference between mortar joints only occurs in a lateral direction; top-to-bottom jointing stays the same. The effect is shown, slightly exaggerated, in 4-1.

Because this differential in lateral spacing of the mortar joint can be quite extreme, very large mortar joints can occur if long log-ends (wide walls) are used with small-diameter buildings. While this situation may not present any structural problems, it will mean: (1) a less attractive wall, (2) the need for mixing a lot more mud, and (3) much more work during the pointing process. For these reasons, I suggest that round or hemispherical buildings of varying outside diameters be limited in their wall

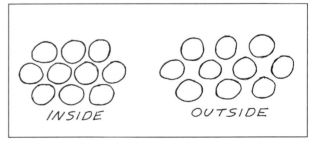

4-1. This is the sort of differential of spacing which will occur on the exterior of a round house.

thickness as per the following chart:

Diameter:	6'	8'	10'	12'	16'	20'	24'	32'	40'	50'
Thickness:	6"	6"	8"	8"	12"	12"	12"	16"	18"	24"

A 6' diameter building? Yes, it is possible. Here is a playhouse called "Littlewood." My son Rohan and his little friends built it up to window height when he was seven years old. Jaki and I finished it up as cold weather set in and little fingers became less nimble. Built in 1983 of "reject" log-ends (considered not good enough for use at Earthwood), the building is still flourishing today and supports an earth roof.

4-2. Littlewood builder Rohan with present-day tenant, Darin.

Here is a tip which will help reduce the disparity between the inner and outer mortar joints: As you handle the log-ends just before placing them in the mud, have a quick look to see which way the log tapers. Many log-ends have a discernible taper, especially white cedar. Always place the wide end to the outside, which helps to equalize the spacing.

ROUND IS EASY

Laying up a round cordwood wall is actually easier than laying up the walls of a rectilinear structure. The primary advantage is that there are no posts or built-up corners to fit the masonry up to; only door and window frames will slow down the masonry work. You can keep the walls round and plumb by following some basic building techniques described after a discussion of foundation options and considerations.

FOUNDATIONS FOR ROUND BUILDINGS

With the round style, cordwood wall construction will commence right on a rigid foundation. The footing width should at least equal the width of the cordwood wall. It is not a good idea to build a 16" thick wall on a 12" wide footing or foundation wall, no matter which of the three styles of building is selected.

In northern climes, attention must be given to potential frost-heaving problems. Water expands about 4 percent when it freezes. There's not much we can do about this law of physics, every bit as constant as the law of gravity. If wet ground freezes under our house or foundation, it will expand and cause an uplifting called "heaving." This can cause structural damage, possibly severe.

There are two different strategies for avoiding frost heaving. The more common method is to build below the depth considered to be the maximum frost penetration in your area. Ask your building inspector for the local figure. In the city of Plattsburgh, NY, just 12 miles from Earthwood, all structures must have foundations extending to at least four feet (4') below ground level. This is also the depth that water pipes are buried to prevent freezing. This strategy is also one of the reasons that basements are still popular. The line of thought is that if you have to go down four feet anyway, you might as well go the extra three feet and have a basement. I am not convinced.

The option of building a "frost wall," a rigid masonry foundation extending down to the local frost depth, is not such a good one with the curved wall style. A track to frost depth must be dug, difficult to do accurately with a backhoe if the house is round. And then comes the difficulty of accurately building a round masonry wall below grade, without the advantage of being able to measure off of the center point. At Earthwood, we built a 16"-thick block wall on the floating slab, and bermed it later with earth, so measuring and keeping round was easy. A round poured wall should be left to experienced professionals.

THE FLOATING SLAB

The best foundation option for the round house, in my opinion, is the floating slab. By this method, a "pad" of good percolating material is built up to a depth of 12" or so above grade, and the concrete slab "floats" on this pad, settling equally as time goes by. Water cannot collect under the slab, because the good percolation characteristics of the pad carry the water to a lower place on site. No water, no heaving. In general, the order of events is as follows. A detailed specific case—Earthwood—appears in Part Two.

(1) Scrape the organic material—grass, topsoil, etc.—off to the side of the site. It will be handy later if an earth roof is desired, or, perhaps, for the creation of a raised bed garden. The scraped area should be six feet (6') in diameter larger than your house; a 46'-diameter pad will be built for a 40'-diameter house, for example. This gives you a three foot (3') sloping skirt all around the house. On a small building, such as my 20'-diameter office, a two-foot (2') skirt has worked well.

(2) Bring in good percolating material to build the 18"-thick "pad" upon which the slab will "float." This material can be coarse (not silty) sand, #2 (1" to 1½" mesh) crushed stone, or river-run (coarse) gravel. Sand is easiest to work with, especially in terms of setting the forms for pouring, but it must be put down and compacted in "runs" (layers) of not more than 6" each. Deeper runs will not compact well. Neither

will dry sand, so we water down each run prior to compacting with a hired power tamper. Crushed stone is pretty much compacted when it falls off the truck. It just needs to be spread to the right depth.

(3) Underfloor drains are installed throughout the pad as per 11-4 in Part Two. If water finds its way under the house by any unexpected means, it is carried away by these underfloor drains to a point downgrade from the home. The drain is made from 4" perforated flexible drain "tile" or tube.

(4) Footings are formed out with ¼" plywood. I recommend a 24"-wide by a full 9"-thick footing. Footings thicker than 9" are a waste of money. The floor should be 4" thick or very close to it. See Figure 4-3. With the sand pad, it is convenient to draw the footing tracks with a hoe, maybe five inches (5") into the 18" pad. The sand which comes out of the tracks can be packed on the outside of the footing forms to help resist against "blow-out" of the concrete during the pour. Insulating the footings is illustrated in Part Two: Earthwood.

With coarse gravel—sand mixed with stones—it is best to proceed as with sand, removing any stones larger than a baseball. Stones bigger than 4" get in the way of setting footing forms and make tamping more difficult.

With crushed stone, with its excellent percolation, a 15" thick pad is sufficient. Spread about 10" of stone, set the footing forms directly upon this 10" layer, and then spread another five-inch (5") run of

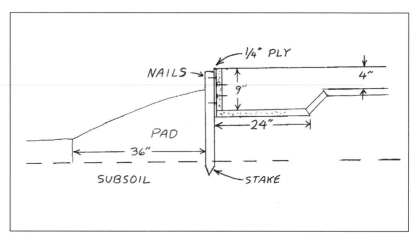

4-3. Cross section of floating slab on its sand pad, also showing shuttering (forming boards) and placement of insulation. In this example, the slab is poured in one piece, unlike Earthwood, where the footings and floor were poured independently.

crushed stone up the outside of the footing forms for support. If a concrete floor is the intent, build the inside of the pad up 5", to the level of the bottom of the floor slab.

(5) Perform all the underfloor requirements detailed in Part Two, such as underfloor plumbing, any needed electrical conduit, and Styrofoam® insulation.

(6) Pour the slab, also detailed in Part Two. With a small building like Mushwood (Chapter 6), the slab can be "monolithic," which means that the footings and floor are poured at the same time. With the monolithic slab, the footings can be thought of as a thickened part of the floor slab. With a large-diameter building like Earthwood, it is better, in my opinion, to pour the footings one day and the floor on a different day.

If you are troubled with expansive clay soil in your area, seek local professional advice before proceeding with the floating slab. Experienced contractors, building inspectors, architects and engineers can all be helpful. A soil engineer or even the county health department can help you to determine if your soil is of the expansive clay variety. Architect Marc Camens has told me of expansive clay conditions he has encountered in Alabama which would have defeated even a floating slab.

FLOATING RING BEAM

I don't know if "floating ring beam" is accepted engineering terminology, so I'd better explain my use of the term. The ring beam is simply a perimeter footing. If there is an intersecting internal footing, this is often referred to as a grade beam. A grade beam can be useful, for example, in cutting down floor joist spans. We used both a ring beam and a grade beam at Log End Sauna, as shown in Chapter Three. In the case of a round house where a wooden floor is desired instead of a concrete floor, the round perimeter ring beam can be "floated" on a pad, just like the floating slab. The walls are really getting the same bearing as with the floating slab, as the floor doesn't contribute much to wall load distribution.

A round central footing is poured at the same time as the ring beam. This footing can be up to 7' in diameter if it is to support a heavy stone mass such as the masonry stove described in Part Two. Whether or not the central footing supports a mass, it provides a bearing surface for the inner ends of a radial floor joist system. If that is its only function, a four foot (4') diameter by 9" thick disk of concrete in the center is sufficient.

The floating ring beam, with its accompanying disk at the center, provides direct bearing for the ends of a radial floor joist system in preparation for a wooden floor. Be sure to cover the pad material with a damp-proof course, such as 6-mil black polythene ("visquene") to stop rising damp. An inch or two of rigid polystyrene insulation is a good idea, too. Keep the floor joists at least three inches clear of anything below them: sand, polythene, rigid insulation, whatever.

This space between floor joists can be extremely useful in running electric lines and plumbing. The outside ends of the joists can be built right into the cordwood walls; the inside ends come together in the center. Plank with 2" by 6" tongue-and-groove planking, as detailed in Part Two. Or use a plywood subfloor and hardwood planking.

Designing joists to carry spans is a complicated matter, even with rectilinear buildings. But with rectilinear buildings, we at least have the advantage of being able to look up span tables for live loads of— say—20, 30 or 40 pounds per square foot; such tables can be found in *Architectural Graphic Standards* by Ramsay and Sleeper, found at any library, as well as many other books on frame house construction. With a round building, such tables do not exist, or, if they do, I haven't seen them.

One of the most important considerations, though, is that bending strength decreases as the *square* of the span. Deflection and "springiness" (the trampoline effect) increase as the square of the span. When you introduce such variables as the kind of wood and its grade, frequency (or spacing) of members, dimensions of the members, a grand piano (concentrated load) ... well, I think you can see the kinds of problem that can be encountered. The most common error of owner-builders in this regard is to design spans too great for the number and size of joists. While I've never actually heard of a floor failing, I have stood on a number of floors that were entirely too springy. I'll return to this subject in Part Two, but, for now, I would strongly suggest the following parameters when designing a radial floor joist system:

When building a round house of up to thirty-two

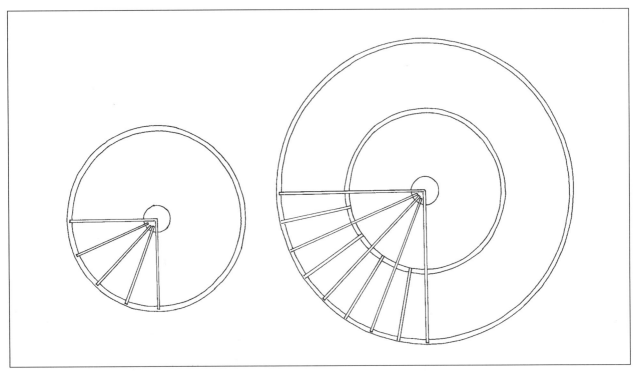

4-4. Round house footings for a radial floor joist system. The single span system on the left would be for buildings up to 32' in diameter. The double span example on the right may be necessary for buildings greater than 32' in diameter.

feet (32') in diameter, it will be possible to proceed with a single ring beam footing and a central footing. This central footing, an 8"-thick disk of concrete, will almost certainly support a center post as well as the inner ends of the joists. As the design gets closer and closer to 32' in diameter, the dimensions of the joists, particularly their height, gets greater. Two-by-tens or two-by-twelves, for example, will be required with the larger diameters by this single-span method, depending on the grade of wood and the number of joists used. With diameters over 32', I recommend an additional ringbeam to support the floor joists halfway. This ringbeam, used to cut the spans in half, is illustrated in diagram 4-4.

Note that the joists join over the intermediate ring beam. Only half of the joists need proceed all the way to the center, because they become more and more "frequent" as they get closer together. As for joist dimensions and frequency, my advice is to have your design checked by an engineer or, perhaps, an engineering class at a nearby college or university. When in doubt, overbuild.

ENERGY NOSEBLEED

Note that with the foundation methods discussed above, the footings are left a clear 4" above the top of the pad. Later, when the home is finished, I like to cover the outside skirt of the pad with 6-mil black polythene and 2" of No. 2 crushed stone. This stops weeds and grass from growing near the cordwood walls, and still keeps the cordwood two inches above a clean surrounding skirt.

To stop direct heat conduction or "energy nose-bleed" at the edge of the house, the exterior of the slab or ring beam should be protected with at least an inch of extruded polystyrene. As this rigid foam will be required under the footing as well, I would advise the use of Dow Styrofoam® Blueboard, the only product I know of with sufficient compression strength (5600 lbs per sq ft with only 10 percent deflection) for use under footings. The exposed edge of the Styrofoam® can be protected from ultraviolet deterioration with surface bonding cement or a coat-

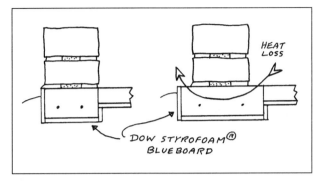

4-5. Left: Cordwood walls the same width as footing. Energy nosebleed is stopped by rigid foam insulation. Right: At Earthwood, there is a potential for a slight "nosebleed" on the south side, but we have not experienced a condensation problem.

ing made for the purpose, such as Retro Technologies' Brush-On Coating. With one-storey buildings, I build the cordwood masonry right to the edge of the footings and chamfer my first mortar joint right over the Styrofoam® as shown in the left side of the diagram above. With a very heavy building, such as a two-storey cordwood home or a building with an earth roof, I go with the broader footing shown in the right side. At Earthwood, we no doubt have a slight energy nosebleed as shown, but only for about 20' or so on the south side of the house. We have experienced no condensation problems as a result.

LAYING OUT THE HOME

With the round style of construction, as well as with the stackwall corners method described in the next chapter, door frames must be set up prior to building the cordwood walls. The installation of these door frames, as well as additional foundation details, appears in Part Two: Earthwood.

It is imperative to lay out several kinds of construction details on the footing or slab prior to proceeding with the cordwood masonry. Besides erecting the door frames in the right places, you will want to know the location of the floor joists (if a wooden floor is to be used), the roof rafters (or second storey floor joists), windows, electrical service entrance, and perhaps even the location of particular design features that you may want to incorporate

into the wall, such as bottle ends and shelves, as discussed in Chapter Seven. The locations should be prominently marked on the concrete itself with a brightly colored wax crayon.

With a rectilinear house, finding these locations is fairly simple: you can scale off your plans and measure from the corners of the building to find the locations of these architectural features. But how do we find these locations on a round house?

My method, described in more detail in Part Two: Earthwood, is to use the center point of the house and eight precise external directional stakes to divide the slab, and therefore the footings, into eight pieces of pie, each exactly the same size. Major components of the house, especially internal walls, follow the cuts between the pie slices. The directional stakes should all be set prior to pouring, if there is to be underfloor plumbing, electrical conduit that rises up into internal walls, or additional pillar footings, such as we employed at Earthwood.

Now, with the circumference of the building divided accurately into eight sections, it is easy to locate all desired design features by transposing measurements from the plans.

This may seem like a small and relatively unimportant matter, but, during the heat of the moment, it is so easy to forget things like windows, shelves, or a special bottle-end design. By having the locations marked on the slab in bright red crayon, the likelihood of building higher than the underside of a planned window frame, for example, is greatly diminished.

CORDWOOD IN THE ROUND

Building curved cordwood walls is basically the same as building straight cordwood walls, already discussed in the previous chapter. The same order of mortar, insulation, and log-ends is maintained. The main difference is that we don't have the advantage of an already plumbed series of panels framed out for us, as with the post-and-beam method, nor can we stretch a mason's line between corners, as is possible with the stackwall system. So how do we keep the walls both plumb and properly curved? Like this:

Clip one end of a measuring tape to a nail driven into a stake at the exact center of the slab. This is the same stake and nail used to set the footing forms. It is retained in the center of the slab as the single most important point in the whole building. With a bright red crayon, and the tape stretched tight, go around the entire building and mark both the inner and outer edges of the cordwood wall on the slab. You will find this easier with two people. One walks around the edge pulling the tape fairly taut, the other uses the crayon. To lessen the chance of error, it is good to tie a brightly colored string to the points on the tape corresponding to the inner and outer radii. For example, at Earthwood, the inner radius—the distance from the center—is exactly 18'. With 16" walls, the outer radius is 19' 4". Although I build from the inside, letting any irregularities in log-end length appear on the outside, I like to see that outer circumference and how it relates to the edge of the footing. Also, it's there if my inner guideline rubs out.

Now, the first course is going to be laid truly round. If we keep the wall plumb from here on in, it will stay round.

You really can't go too far wrong for the first two feet of height, nothing that can't be easily corrected with succeeding courses. After about two feet of height, I begin to use my four-foot level frequently to make sure the wall is going up straight. I use the plumbing bubble, mostly, although I will occasionally check the actual level of the log-end, too. My rule of thumb is to use the level to check the wall for plumb each time I lay up a large log-end. Typically, this is about every six or eight pieces. As there is bound to be some irregularity in the cuts of the log-ends, and previous work has not always been laid up perfectly, the level must be used in such a way as to show you the average plumb over its full height.

I use my inner red crayon line as the base for checking the plumb of the wall as long as I can, but higher up the wall, it is necessary to use another log-end known to be correct as the benchmark for further work. Other options: use the level in combination with a straight two-by-four stud to show

4-6. (left) Jaki can see that this round cordwood wall, for all its irregularites, is (on the average) plumb. 4-7. (right) Care was taken to make this high window frame plumb, which then gives an excellent guide for making the masonry plumb.

4-8. Cheaters help fill the large mortar joint near doors and windows on the outside of a small round building.

plumb, or buy a mason's six-foot level, which is quite expensive.

Frequent use of the level, in combination with an accurate base line on the footing, will assure that the wall has the right curve and is vertical. If there is a place where the wall has wandered an inch out of plumb, that's nothing to worry about. Slowly bring it back over the next two or three courses. Do not attempt to make the entire correction on the very next course. Take special care in plumbing—and leveling—window frames. Not only will they look good, but they will serve as excellent benchmarks to use in plumbing nearby cordwood masonry. More on window frames in the next chapter.

On small buildings, such as Rohan's shed shown in Figure 4-8, the difference in lateral width near door and window frames can be quite extreme. The door frame is like a huge log-end and the parallel edges are not aiming for the center of the building. We add small "cheaters" to the outer mortar joint.

These cheaters are small wooden pieces—wedges work best and they actually help to tighten up the large expanse of mortar while making the area more pleasing to the eye.

ON PARALLEL RAFTERS

Some builders of round houses have used parallel rafter systems, others have used a "hip" roof design, again so that rafters can be kept parallel. The main purpose of these efforts seems to have been to accommodate the use of fibreglass insulation and sheet rock ceilings. The down side is that these ceilings hide a potentially beautiful support structure, and the finishing work is difficult, especially with the hip design. My final complaint with the parallel system is the weakness of the support structure for the planking at the edge of the building. Donald Barth, who built a beautiful little round vacation cabin with a hip roof in central New York, attended to this situation by the following five step process: (1) Frame the four-sided hip roof over the round cabin, with double rafters for the four long diagonals; (2) Trim all the parallel rafters (and the four heavy diagonals) the same distance from the center, giving about a 15" overhang around the building; (3) Nail a one-by-eight fascia board all around the building, joining all the ends of the rafters; (4) Nail the planking to the rafters and also to the fascia board; (5) Trim the planking flush with the outside of the fascia, using a chainsaw, reciprocating saw, or jigsaw.

THE RADIAL RAFTER SYSTEM

I feel strongly that the radial rafter system, by which all the rafters aim for the center of the building is (A) easier and (B) more attractive. Cost-wise, I doubt if there is much difference one way or the other.

With the radial system, which will be detailed in Part Two: Earthwood, the heavy timbers are exposed. I like Charles Wing's comment on exposed timbers:

> I, too, need to see what holds my roof up. I never walk under a ladder without looking up; how can I be expected to feel secure under an expanse of plaster, held up for all I know by levitation or three horsehairs and a drywall screw? No, in order to sleep soundly I need to know that I'm being protected by a band of stalwart rafters overhead.[15]

This confidence is particularly important when there is over 100 tons of wet earth and snow overhead, as at Earthwood. Visitors have expressed a strong sense of comfort and security from seeing the pleasingly proportioned five-by-ten rafters overhead. I've appreciated beautiful heavy beams ever since my early pub-crawling days in the south of England.

With the radial roof, the roof membrane is placed directly on the wooden deck, typically two-by-six tongue-and-groove planking. Next comes the rigid foam insulation, which protects the waterproofing membrane from sunlight (ultraviolet radiation) and freeze-thaw cycling, any roof's two worst enemies. Then comes a cheap insurance layer of black polythene, 10-mil. if you can get it. The roof can be finished with 3" of crushed stone, to stop the plastic and insulation from blowing away, or, if the proper engineering has been worked out, the best roof—the earth roof—can be applied. More on this in the second part.

THE PLATES

The rafters should not bear directly on the cordwood walls without benefit of a "plate" to distribute the load, although it has been done. My strong view is that if a rafter is simply laid up in the wall like another log-end, the concentrated load of that rafter, not to mention the impact load of second floor joists, places undue strain on a cordwood wall.

At Earthwood, we used a very powerful plate system which, in combination with almost continuous window lintels, created a virtually continuous tensile ring at both the second floor and roof levels. This is described in detail in Chapter 14.

For most buildings, however, I believe that a simple wooden plate of 10" to 12" in width at each joist or rafter location is quite sufficient. Both my earth-roofed office and our dome-topped Mushwood cottage make use of such a simple plate system. In these buildings, both of which have 12" cordwood walls, I cut the sixteen plates from two eight-foot two-by-twelve planks, as per 4-9. I put a few roofing nails in the bottom of each plate before laying it up in the wall, to lessen its chances of getting knocked loose during rafter installation.

But in addition to better bearing strength, there is another compelling reason to use the plates: It is much easier to set the second floor joists or rafters once the plates are in place. And the plates are easy to set in the right place if (A) their location has been marked on the foundation with a bright crayon and (B) you employ an "idiot stick" to get them at exactly the right elevation. An *idiot stick?* Yes, it is my own invention, and makes the job almost foolproof for people like… me.

My stick is made from a straight two-by-four stud, with a dull point shaped at the bottom end. The point assures a more accurate measure, as there is no danger of riding up on one edge of the stick or the other while holding the stick up to the wall. On the stick, I show in graphic illustration the location of the rafter, the plate, the last mortar joint before the plate, and even the last course of log-ends. The stick tells me right away if I've laid a log-end up too close to the underside of the plate. If I have, I can remove it immediately and replace it with a smaller one, or, perhaps, split the appropriate amount off the top with an axe and hammer.

There are no numbers on the stick. An idiot can use it. A tape measure is an awkward tool to use for this purpose and there is too much chance of making a mistake, getting numbers mixed up. The idiot stick can also be used to set the bottoms of window frames all at the right height. Just mark on the stick the window frame, the last mortar joint, and the

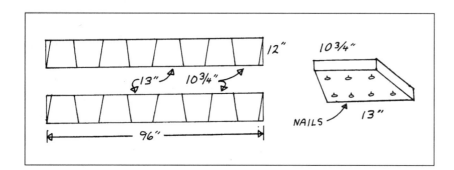

4-9. These simple wooden plates help distribute rafter load and make setting the rafters much easier.

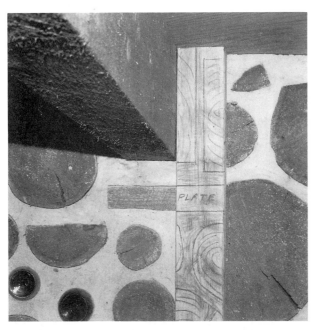

4-10. The top of the "idiot stick" shows the location of the plate, the last mortar joint, and the maximum height of any log-end beneath the plate.

maximum allowable elevation of the last log-ends before the frame is set. Usually, there will be quite a number of windows of the same size in a building and looks bad if they are not all at the same elevation. If the stick is made carefully, checking and rechecking the plate and window frame heights, it will be one of the most valuable tools you will have on the job.

Installing the roof itself on a round house is illustrated in Part Two.

COUNTERACTING OUTWARD THRUST

Don't worry. This little section is not nearly as dry as the title implies. In fact, we can have some fun with an experiment suggested by the redoubtable Jack Henstridge, one which it is perfectly safe to try at home.

I cut a piece of copy paper lengthwise in half and taped the top ends to the bottom ends, one to make a cylinder and one to make a cube. I roofed the models with cardboard beer coasters. I carefully placed a five-pound brick on the roof of each structure, a tremendous load considering the flimsy walls. The round building held. The square one crushed immediately; I could not fully release all of the weight of the brick. The experiment was repeated with the same results.

Any square people left?

Now, this little test shows one kind of structural integrity, but paper, made of thousands of interlocking fibres, is probably much stronger on tension than cordwood, by unit width. We must still be wary of outward thrust.

With a radial rafter system, each rafter bearing on a plate, the roof thrust will still be outward if the ends meeting at the center are unsupported and not tied together. There needs to be a strong compres-

4-11. (left) Little paper houses with beer coaster roofs... 4-12. (middle) A five-pound brick is placed carefully on each...
4-13. (right) Any questions?

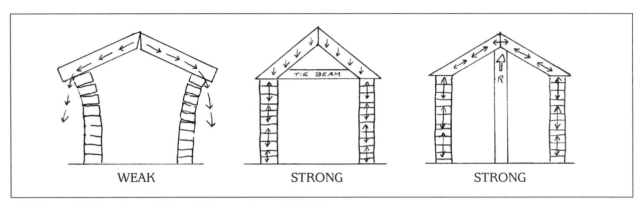

| WEAK | STRONG | STRONG |

4-14. Arrows indicate lines of thurst. "R" = reactionary load of post.

sion member at the center, such as a post or a stone mass. Several center support methods are illustrated in Part Two: Earthwood.

The steeper the roof pitch, the more important this concern becomes. Twelve-in-twelve (45°) roof pitches will put an awful outward thrust on the wall unless held from spreading by use of tie beams.

Many students at Earthwood want to build a two-storey round house with a considerable portion of the second storey space devoted to a cathedral ceiling. Leaving aside the difficulties with heating such a space, let us consider structure: The tall two-storey cordwood wall starts to encounter a problem with what engineers call "slenderness ratio," the relationship between the height and the width of a load-bearing structural component: post, wall, whatever. If the open space is an absolute must for spiritual well-being, at least incorporate the use of a few radial tie beams flying through the space from the outer wall to the center. They actually look great. Architecturally, they call even more attention to the overhead space. And you can hang iron kettles, beer mugs, or knick knacks on them. Finally, you can sleep easier at night.

OTHER SHAPES

Other curved wall shapes can be built besides truly round. Cordwood masonry is a very flexible medium. But when the builder departs from a round structure (or true hemisphere) and begins to work with spiral shapes, ellipses, or free-form curved walls, he or she departs from the builder's way of thinking and enters

the realm of the artist. Actually, building the cordwood walls of an eccentric shape would probably not be any more difficult than a true round, once the slab is in and the basic shape is marked out in crayon. Walls should still go up on a true vertical, easy enough to do with frequent use of a four-foot level. The real challenge comes when it comes time to put the roof on. Neither radial rafters nor parallel rafters will work symetrically. Elements of both might be necessary. This is a good time to heed the advice of Jack Henstridge: "Build a model. If you can't build the model, don't even start the house."

4-15. This model, made to the scale of Star Trek figures, combines elements of both Log End Cave and Earthwood. Jointing details can be worked out on such a model. Now let's beam over to Built-Up Corners.

5

Built-Up Corners

OVERVIEW

With the built-up corners method, the cordwood masonry itself does all the load supporting. Corners are built first so that a mason's line can be stretched between them to assist in keeping the walls plumb.

There are several ways to build up the corners. The Northern Housing Committee (N.H.C.) of the University of Manitoba suggests laying wood blocks with mortar, as with the rest of the wall. They build cordwood walls 2' thick to deal with the heating requirements of homes in the Canadian West. Their

corner blocks, then, are 30"-long eight-by-eights, milled on three sides.

With three sides finished, the corner will be strong and the outside edge is handy for attaching a corner clip for holding the mason's line. Malcolm Miller, of Fredericton, New Brunswick, built his corners with full-size six-by-six timbers, pressure-treated for extra-long life. Rough-cut six-by-sixes are quite commonly found in building supply yards. Given a choice, the builder should try to obtain the driest ones.

Even though the bond between wood and mortar is not very great, these mortared-up corners are

5-1. Stackwall barn in Hemmingford, Quebec, Canada, built in 1952.

5-2. Corner detail of the Hemmingford barn.

incredibly strong. In 1980, I visited a stackwall barn in Hemmingford, Quebec, built in 1952. The structure was built with 10" cedar log-ends that had not been barked. The corners were built with the same type and length of log-end as the wall, 3" to 4" in diameter. Like the log-ends, the corner pieces were cylinders for the most part. The structure was built without any special care taken to keep log-ends from touching one another and there is a dead air space in the mortar matrix by way of a thermal break. All in all, I would not have expected this building to have lasted as long as it had—28 years—but, after inspecting the barn I was convinced that the walls would still be sound for another 28 years…and then some. In other words, a corner built with short, round, unbarked cedar log-ends, some of which are actually touching each other, will last at least a half-century. If care is taken, such as exercised by Malcolm Miller and the N.H.C. in their structures, the buildings will last hundreds of years, assuming a good foundation and periodic roof maintenance.

I offer three alternatives to folks who are unable to obtain seasoned, squared timbers for the corners:

(1) Use "beam-ends" cut from old barn beams in the built-up corners. (Note: The use of creosote on externally exposed barn beams will give a pleasing contrast to the cordwood masonry and preserve those old timbers for a lot longer. Creosote should *never* be used indoors, as the obnoxious smell will persist for years.) (2) Spike green timbers together with 10" spikes, the way many people are building "traditional" log cabins these days. Figure 1-2 shows a barn in Wisconsin that looks as if it may have spiked rather than mortared corners. (3) Scrounge some dry recycled two-by-sixes and lay them up as shown in Fig 5-3. This drawing assumes a wall thickness of 16". Betsy Pugh employed this method in her house, shown in Chapter 6.

No matter what material is used in the built-up corners, the technique for laying up the walls is the same. The corners are built up about 2' or 3', then a mason's line is stretched between two corners and the cordwood masonry walls are infilled by methods similar to those described in Chapter Three. Obviously, it is very important that the corners be built plumb in both directions. A 4' level is the best tool for this purpose.

Although the barn in Hemmingford and many others like it were successfully constructed with walls 10" thick, Jack Henstridge and I agree that 12" should be considered the minimum thickness for a

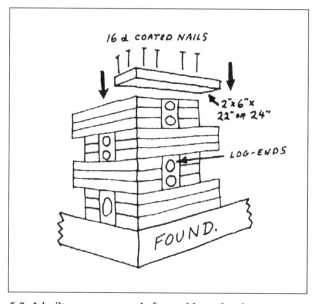

5-3. A built-up corner made from old two-by-sixes.

load-supporting wall in a dwelling. Sixteen-inch walls are preferred, especially in cold climates. A 24" wall is 50 percent better in terms of insulation, but is not strictly necessary for load-supporting.

FOUNDATION OPTIONS

A frost wall is easier to do with a rectilinear structure, so it becomes a somewhat more viable option than it is with the round style. With a 12" wide cordwood wall, the frost wall could be built of 12" wide concrete blocks, as we used at Log End Cave. If 16" thick walls are to be used, the frost wall could be built of 16" long concrete corner blocks, laid widthwise in the wall-like log-ends. We used this technique for our below-grade walls at Earthwood. With 24" walls, I'd stay away from the frost wall or basement techniques altogether, in favor of the floating slab discussed in the previous chapter.

In fact, I have to say that my personal preference for economy, frost-heaving protection, and ease of building is the floating slab, even with stackwall corners. The technique is the same as described for the round slab in the last chapter, and further illustrated in Part Two. The option of the floating ring beam is also there, in case a wooden floor is preferred. A grade beam (or two or three) can be used to shorten floor joist spans, as per 3-39.

The footings and floor can be poured at the same time—remember the "monolithic slab" in the last chapter?—providing that external dimensions do not exceed 40'. If either dimension exceeds 40', the footings should be poured first and the floor should be poured a few days later with an expansion joint between the floor and the footings. The expansion joint can be homosote board made for the purpose, or ½" or 1" extruded polystyrene insulation. My recommendation for footing dimensions is as follows:

Wall	Footing Dimensions	
Thickness	Height (H)	Width (W)
12"	8"	16"
16"	9"	16"–24"
24"	9"	24"–32"

There is no gain in making footings deeper than 9". Additional concrete is better used in making the footing wider, thus spreading the weight load over a

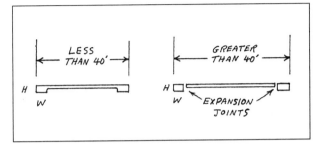

5-4. If external dimensions exceed 40 feet, the footing should definitely be poured first, and the floor later. Wider footing dimensions given are for two-storey buildings, buildings with earth roofs, or buildings in areas where the subsoil is known to have a lesser bearing strength.

greater area. Half-inch rebars should be wired into the footing, supported about halfway between the top and bottom edges of the forms, or slightly lower. Keep 8" to 10" between rebars.

The floating slab thus created offers several advantages over other foundations in areas with deep frost conditions. First, the pad of percolating material is the best protection against frost heaving. Second, much less concrete is used than on foundations requiring excavations to a 4' depth or more. Third, it is fast and easy, and well within the capabilities of the owner-builder. Finally, the insulated floor acts as a thermal mass, helping to maintain a more constant temperature during the winter.

I know that some people have an aversion to concrete floors—I'm quite fond of wooden floors myself—but I think that a good part of this aversion stems from a presumption that concrete floors are cold and damp, probably because basement floors often display these characteristics. If insulated and built on a percolating pad as described, a concrete floor is neither cold nor damp. Carpets, vinyl and parquet flooring can all be installed on a good smooth concrete floor, although some of the value as thermal mass may be lost with a thick plush carpet.

Besides the ring beam and grade beam option, in which floor joists are installed, there is another way to install a wooden floor. Furring strips are nailed to the concrete floor slab, and a wooden floor is fastened to the furring strips. Two-by-four stock would make good furring strips for this purpose. I'd still use an inch of Dow Blueboard beneath the slab, as a guard against condensation. The main disadvantage of this method is that we don't gain the chase ways for installing plumbing and electricity, as we do with the ring beam option.

WINDOWS AND DOORS

A bonus with walls that are 16" to 24" thick is the wide window wells, ideal for sitting, growing indoor plants, or starting garden seedlings. Windows for a 16" wide wall can be framed as shown in 5-5. The tie pieces—also called "scabs"—are made of scrap one-by-six material. They not only tie the inner two-by-eight frame to the outer frame, but they key the whole frame into the cordwood wall. The scab occupies the insulated cavity close to the window frame. The diagonal temporary brace is only installed after the frame has been checked with a carpenter's framing square. It is removed after the cordwood walls are completed.

Door frames can be constructed in much the

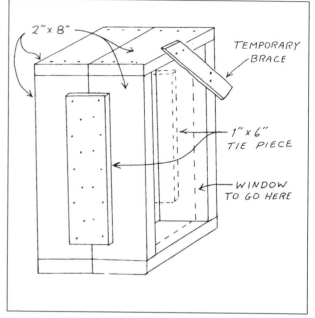

5-5. Window framing detail.

SETTING A WINDOW

5-6A. First, a flat area of mud is created at the right height and at the right location.

5-6B. (left) I set the frame onto the flat bed of mud. The hammer is used to gently tap it in, plumb and level.
5-6C. (right) It's plumb...

5-6D. (left) ... and it's level! 5-6E. (right) The window frame is held square by the diagonal brace. Note the wooden "key," which locks into the insulated mortar joint. Cordwood is built up to both sides of the frame right away, to help hold it

same way. I always use door frames at least 4" thick, so that they don't bow, twist or warp. I also hang some pretty heavy homemade doors with large strap hinges, another reason for thick frames. Or, the builder might change to the post-and-beam style just to frame the doors, as shown in 5-8.

With any style of cordwood construction, door frames must be erected before masonry work can proceed. They can be fastened to the footings or floating slab with metal pins. As this is an important and often misunderstood part of the construction, let us examine the process in some detail.

INSTALLING THE DOOR FRAME

The door frame must be the same width as the cordwood wall. Figure 5-7 shows the kind of frame we used at Earthwood: two sets of four-by-eight framing standing side by side to make up the 16" width of our wall. With a 24" thick stackwall building, you might use three four-by-eight frames scabbed together, or even four six-by-six frames.

My eleven-step process for setting up the frames is quite straightforward, strong, and much easier than it sounds.

(1) Mark the location of the components on the slab or footing, using a bright crayon. The space between the frames must make sense with the homemade or prehung door which you have in mind. With thick doors, such as the 4" thick Earthwood door, allow a full ½" extra width to accommodate for the swing of the door. A 36" door which is 4" thick, for example, requires 36½" of space to swing. Or, the doorway can be exactly 36" wide, and the door can be made 35½" With a bought prehung unit, such as a Stanley or Therma-Tru door, use the manufacturer's rough opening dimensions.

(2) Cut all the components of the frame. The heights of the pieces will correspond to either a homemade door which you are planning to build, or to the height of a manufactured prehung unit. Remember to give the homemade unit a little clearance to swing. Follow the manufacturer's instructions for rough opening on bought units. The lintel will be the outside width of the door plus the amount extra you want to tie into the wall. At Earthwood, our front door opening is 38". To this we add 8" for the two 4"-wide side frames, and another

8" because our lintel extends an extra 4" on each side. All told, the lintels are 54" long .

(3) With a half-inch (½") masonry bit, drill a hole at the center location of each of the four four-by-eight door posts. The depths of the holes should be the same or just slightly more than the depths of leaded "expansion shields" which are available at most hardware stores. Usually, these small leaded receiving sockets are about 1½" long. The expansion shield will give the drill size and lag screw size right on it, such as ½D–⅜S (½" drill, ⅜" lag screw.) After digging or blowing out the masonry dust from the holes—close your eyes—the expansion shields can be hammered into the hole, flush with the footing or slab.

(4) Install four hex-headed lag screws (probably a ⅜" shank) into the leaded expansion shields until they are tight. Cut the hex heads off with a hacksaw, leaving a good inch of the ⅜" screw sticking

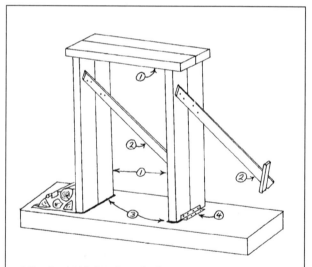

1. Four-by-eight timbers frame the door.

2. Temporary support braces plumb and steadies the door frame.

3. Damp-proofing course

4. 2" x 2" x ¼" angle iron fastened to footing with expansion bolts and to frame with lag screws.

5-7. An angle-iron bracket is shown anchored to the foundation with lag screws fastened into leaded expansion shields. This adaptation of the process described above was used successfully at the home of Bob and Aggie Evans, shown in the next chapter.

out of the slab. These are the positioning pins.

I'm sure it will have occurred to some readers that anchor pins could have been inserted into the footing or slab prior to the concrete setting. This is not usually successful. For one thing, it is very hard to do a smooth flat job of troweling near such anchor pins, so the base is irritatingly irregular. And Murphy's Law says that they won't be in the right place anyway. The masonry drill, expansion shield, lag screw method is easy and accurate. Trust me.

(5) Install a "damp-proof course" on the concrete, made of rectangles of asphalt shingle or 90-lb roll roofing. The pieces should be the same size as the bottom of the door frames, 4" by 16" in this case. Holes will need to be cut in the asphalt to set them over the positioning pins. I have also used Bituthene Waterproofing Membrane for this purpose. Besides stopping what the British call "rising damp" into the door frame, the thick asphalt material also helps to steady the post when it is stood up

(6) Scab the sides of the frames together using one-by-six boards. From here on, you will need at least two people to complete the job.

(7) Stand one side of the frame up on the two positioning pins, so that all edges are exactly over the crayon marks on the slab. A good trick to make this easier is to place a couple of concrete block guides on the slab right at the edge of the marks. With the frame held in place and plumb, climb a stepladder and, with a heavy hammer, give a good belt on the top of both the inner and outer pieces (jambs) of the

frame, so that an impression is made on the bottom of each jamb.

(8) Ease the heavy frame back down. With a ⅜" bit, drill holes into the bottom of the frame, just a little deeper than the height of the positioning pins.

(9) Set the frame up again on the positioning pins and brace it with a long diagonal nailed to the frame and staked to the ground. Set up both jambs in the same way.

(10) Mark the underside of the lintel pieces with two pencil lines, corresponding to the width of the doorway as it actually is at the bottom. Leave the same amount of wood showing on the outside of the pencil lines, that is: center the width of the doorway on the lintel pieces. Now, haul the pieces up the stepladder and install them with 8" spikes. Predrilling holes for the spikes—or long lag screws—is advisable.

(11) Plumb, square, and brace the whole unit, using diagonal temporary bracing, as shown with the windows.

LINTELS AND SILLS

The N.H.C. recommends lintels over doors and windows, and sills beneath windows. The lintels can be a pair of six-by-six timbers installed just above the window or door frame. One of the lintels should be kept flush with the interior surface of the wall, the other flush with the exterior of the

5-8. Door framed with heavy posts.

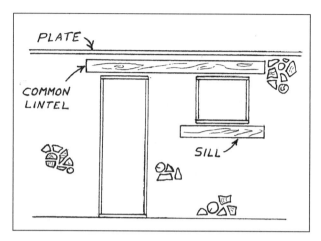

5-9. Lintels and sills.

wall. Thus, they are exposed and are a pleasing design feature on both masonry surfaces. The lintels should extend 6" to 8" into the masonry on each side of the door or window. Their purpose is to pass the weight load of the roof onto the masonry walls, without placing undue pressure on the window or door frames. The sills that the N.H.C. recommends beneath the windows are built in much the same way as the lintels (see 5-9).

The positioning of the window frames in the cordwood wall should be carefully figured. Room

should be allowed above the window frame for a 4"-thick lintel and a 2"-thick plate. (The plate system is described in detail in Chapter 14). The heights of the plates will be determined by the door frames, in most cases, and the positioning of the windows will be determined by figuring down from the plates. Figure 5-10 shows how to determine how high the cordwood wall should be before a 4'-high window frame is installed. No log-end in the last course below the window frame should be within ¾" of this calculated height. The window frame is simply set in the mortar at the right height, plumbed, and supported by temporary braces. We generally try to lend some stability and support to the window frame by laying cordwood masonry a foot or so up each side of it on the day of installation. Make sure the windows themselves fit the frames before laying cordwood. The frames must be accurately squared with temporary bracing.

Some builders have used neither lintels nor sills with the windows of the house. I have left them out of small buildings myself, like Mushwood and my office at Earthwood. But when I do leave the lintels out, I make sure that floor joists or rafters do not bear down directly on the middle of a two-by-two window frame. In other words, windows are planned to fall between rafters. With the heavy Earthwood house itself, we used 4"-thick lintels, made from two

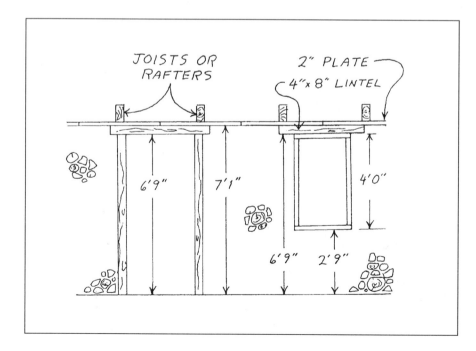

5-10. Determining the location of the window frame in a masonry wall.

adjacent pieces of four-by-eight timber, but we did not use the sills, and have suffered no ill effects from leaving them out. Only a few cordwood builders that I'm aware of have used the sills. They look nice and probably help tie the whole wall together, but they involve more work and planning, as well as more dimensional material, which must usually be purchased. Malcolm Miller used the sills very effectively in his beautiful home in Fredericton, New Brunswick.

5-11. A corner made up with quarter-round quoins.

SELECTION OF QUOINS

Webster's Unabridged defines *quoin* as "any of the large squared stones by which the corner of a building is marked." While strictly a stone masonry term, I can think of none better to describe the special log-ends used to make up a stackwall corner. There are many choices when it comes to selecting the type of quoin to use in the built-up corners.

Many builders, in the past as well as in more recent times, have used nothing more elaborate than round log-ends, usually a little longer than the pieces used for the rest of the house. The house of Irene Boire and Elizabeth Mangle, one of the case studies in the next chapter, is a better example of this than the Hemmingford barn already shown at the beginning of this chapter. For many builders without access to any form of milled wooden blocks, this may be the only viable or economic option.

The round quoins, however, would not be my first choice. My view is that they are not as stable in their bed of mortar as a squared piece. That is, there is not a good broad bearing surface of mortar to distribute the load of the round quoin itself. Similarly, it is difficult to lay a good bed of mortar on top of such quoins; the mud wants to slide off the rounds.

Another option which has been used successfully by cordwood builders is to make quoins of quartered logs. The pieces can either be split from relatively clear-grained logs, or they can be quarter-sawn at a sawmill. They are laid up with a flat side down and the other flat side out, as shown in 5-11.

There is no chemical bond between wood and mortar, just friction. And there is the possibility of an outward thrust on the quartered quoins. (Think of the mortar acting as a wedge trying to drive the quoins outward.) For safety and peace of mind, I would advise the use of roofing nails set into such

quarter-round quoins, their heads sticking up about a half inch (½') to grab the mortar joint. By this method, the mortar and quoins are more truly joined. This technique with the roofing nails may be employed with any type of stackwall corner, of course, but I see it as particularly valuable with the quartered quoins.

There are at least two homes with highly unusual built-up corners. Jaki and I visited one in North Carolina having stone corners; the other, in Ontario, has concrete block corners. In both cases, the corners were quite attractive. You had to look twice to see that the stone corners weren't cordwood (or that the cordwood walls weren't stone) as both sections of wall had that deeply recessed pointing typical of the Blue Ridge Mountain area.

The concrete corners were a complete surprise. A student at one of our cordwood classes brought photos of several cordwood buildings which I had not seen before. One used what appeared to be 12" concrete corner blocks as quoins. The real surprise was that this looked very nice indeed. It reminded me of the Victorian stone houses in Scotland which made use of cut sandstone quoins in the corners and around windows, and rougher granite stonework along the sidewalls.

It strikes me as an important consideration to address heat loss in masonry corners of block, brick, or stone. Some sort of thermal break should be incorporated into the corner. I'll leave any readers interested in nonwooden corners to work out the problem of heat loss for themselves. Call me conservative, but I'll stick to wood masonry corners.

SQUARED QUOINS

We've illustrated and talked about round, quarter-round, stone, and block quoins. But the best quoin, in my opinion, is made from squared wooden blocks: four-by-eights, six-by-sixes, four-by-sixes, and the like. Just as good are logs squared on two parallel sides or squared on three sides. Such logs can sometimes be found at a reasonable price where horizontal cabin logs are milled. There may be some lying around the yard which are too short to sell, or, perhaps, have a very slight bend in them which makes them unusable in a long cabin wall, but quite satisfactory when cut into 24" or 32" lengths.

Incidentally, cuttings left over from a cedar log cabin, milled on two or three sides, are an excellent source of good log-ends as well as corner pieces. Keep your eyes peeled for horizontal cabins going up in your area. There are always lots of pieces left over. You could even offer to trade good firewood for the cedar pieces, an offer the builders—now your new friends—will probably jump at.

INTERLUDE

This reminds me of a true story that goes way back to my days as seminar leader for Earth Sheltered Housing at the Mother Earth News Eco-Village. Jack Henstridge was the Cordwood Masonry instructor. John Shuttleworth, founder of *The Mother Earth News*—the instructors affectionately called him the Big Mother—came to our morning meeting some-what flustered. The fellow who was supposed to teach the Log Building workshop didn't show up, must have got lost on the way to North Carolina.

"Look, I've got 20 people signed up for Log Building this afternoon," says John. "Can any of you guys teach a class like that? How about you, Jack? You're free this afternoon."

Jack said he'd have a go. After all, he almost built a log house just before he built his cordwood place. That evening, over a beer at the motel, I asked Jack how his Log Building seminar went.

"It was great, Rob!" came the reply. "I converted 90 percent of 'em to cordwood!"

MATERIALS LIST

Once you've found a source of good corner material, it is well worthwhile to make up an actual materials list of pieces required, and how many of each length. Let's walk through a simple example to give you the idea:

I've decided on six-by-six squared timbers, and 16"-thick cordwood walls . My house has four corners and each one is 8' 2" (98") high . Each course will raise the corner up about 7"...remember the one-inch mortar joint. 98 divided by 7 equals...let's see...exactly 14 courses. (I chose numbers to make it come out even, not a bad idea in real life.) So there will be seven layers of quoins on each side of the corner. On a 16" wall, I'm inclined to go with 24" pieces alternating with 32" pieces, as shown in 1-6. There is an advantage in strength by alternating quoin length, which will be explained a few pages hence. Stay tuned.

So, for each corner, I need 28 quoins (2 times 14 courses). Half of them will be 24" long, half will be 32" long, or 14 of each. There are four corners, so I need 56 quoins of each length to build the corners if this houses 56 at 24" and 56 at 32" . If I were buying 8' squared timbers from a sawmill or building supply yard, I could get four short pieces from each eight-footer with no waste . I'd need 14 pieces to make my shorts. I can get exactly three of the long quoins from each piece (3 times 32" = 96" or 8'). Nineteen pieces will give me 57 long quoins, one to spare. I can make all my corners out of 33 eight-foot six-by-six timbers.

Don't forget to make similar kinds of calculations for any needed plates, lintels or sills. It is a good idea to get all this dimensional timber on site as early in the game as possible.

Not only will this gain valuable drying time, but you won't have to worry about running out during construction, which throws the production schedule all out of whack.

BUILDING THE STACKWALL CORNERS

Welcome to Cordwood 202. Stackwall corners are more difficult to build than a straight section of wall, and they really have to be done first, so that they can serve the very useful purpose of acting as a guide for stretching a mason's line from corner to corner, which makes the long runs of cordwood wall that much easier and more truly plumb and straight. It takes a mason quite a while to become a "corner man," qualified to build brick or block corners on a skyscraper. And he gets paid more for his extra skill. But you haven't time to get qualified. You have to jump in at the deep end.

I'm not trying to scare anyone away from building stackwall corners. They are not that difficult. Just proceed slowly, methodically, and carefully. Use that four-foot level on each piece, checking for both plumb and level.

Have all the quoins and any other needed log-ends, such as a certain size of "filler" piece, all ready and stacked near the work site. Examples of fillers can be seen in Figs. 5-3 and 6-35. I always build a dry-stacked mock-up of the corner with the very pieces which I am going to use. This takes a lot of the figuring out of the building process itself, so that I can concentrate on the quality and the elimination of "energy nosebleed."

STACKWALL CORNERS AND "ENERGY NOSEBLEED"

If built correctly, there will be no place in the stack-wall corner where the outer mortar matrix joins directly with the inner mortar matrix. No direct wick for the transfer of heat. No *energy nosebleed*. Avoiding a direct mortar join does take a little thought and concentration, but becomes second nature after a while. You'll know you've made a mistake when you come to pour in the insulation and find a mortar bridge in the way of the insulation trough. How did that get there? Simply cut out the offending mud with your gloved hands or trowel, and proceed.

The sequence of step-by-step pictures on pages 84 and 85 shows the order of events. The corner shown is one which might be used where a wall is to be turned for strength and terminated, such as where a garage door frame meets a side wall. In a rectangular house, the cordwood walls would continue in each direction, as seen in 1-6 and 5-18. Note that an insu-lated space separates the two mortar matrices throughout the corner. Only at the end of this parapet wall do the inner and outer mortar joints finally meet.

It is obvious that the direction of the quoins alter-nates with each course: if the first pieces run east-west, for example, then the quoins of the second course will run north-south. What is less obvious is that the quoins of the third course are of a different length than the first course. Remember the materials list having two different sizes of quoin? In 1-6 it is clear that the third course quoin are about 6" shorter than those of the first course. The fifth course is back to full length. The courses are deliberately staggered so that the stackwall corners tie in with the long cordwood walls better. If all the quoins are of the same length, there is the danger of a vertical shear crack forming right along the line of the quoins. A good example of a strong corner is shown in Figs. 5-18, 5-19, and 5-20 on pages 85 and 86.

PLATES

As the weight of the wall puts an outward stress on the corners, it is important that the corners are tied together with plates. The plates are also necessary for fastening rafter or floor-joist systems. They pro-vide a flat surface upon which to work and dis-tribute the momentary load over the cordwood wall. Without the plate, rafters with a heavy load will be trying to "slice" into the wall. If it's not pos-sible to span from corner to corner with a single plate, it is advisable to nail a double plate together in two courses. Plate material can be rough-cut or planed two-by-fours or two-by-sixes and they should be spiked into the built-up corners. The N.H.C. recommends that 5" spikes be driven into the plate every 2', and the ends bent on the under-side to act like anchor bolts. They advise that the plates be set into 2" or 3" of mortar and then tapped level. Figure 5-22 shows the corner detail of a strong plate system.

Note that there is an inner plate and an outer plate. If two-by-twelves are used on a 24" wall, or two-by-tens on a 16" wall, the inner plate may be omitted, though the interior appearance will not be as "finished." On a 12" wall, two-by-twelve plates would be excellent; they'd be strong and would fin-ish off the top of the wall beautifully. The owner-builder should make the best use of the available materials in deciding what plate system to employ. Cost and the degree of interior finish desired are two

STACKWALL CORNERS: STEP-BY-STEP

5-12. First course, mud and insulation.

5-13. The first quoins are set.

5-14. (left) Second course, mud again, then insulation. No "energy nosebleeds." l5-15. (right) Checking the second course quoins for level.

5-16. (above) The second course quoins run the opposite direction to those of the first course. 5-17. (right) Keep checking plumb with the four-foot level.

GOOD EXAMPLES OF STRONG CORNERS

5-18. An outside corner at Jerry Mitchell's house.

5-19. Much less of the quoins shows on an inside corner.

5-20. (left) The top of one of Jerry's corners. 5-21. (right) The main part of the house opens into a kitchen addition at Jerry's house. Note the use of the "Mitchell posts" halfway along each wall, also seen in 5-18.

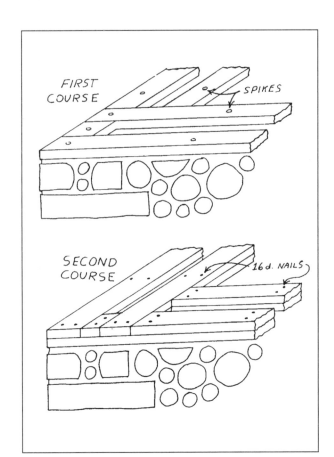

5-22. A strong plate system.

important determining factors to consider. I tend to "overbuild" slightly when making this kind of decision. Low construction cost and structural integrity are not mutually exclusive in an owner-built house.

An interesting innovation at the newly constructed Mitchell home is the use of two six-by-six posts halfway along the walls of the house. These posts can be seen in the previous pictures, as well as in the elevated view of 5-21. I call them the "Mitchell posts." Their purpose is to provide an intermediate location to join Jerry's heavy six-by-six plate beams together, just before floor joist height. These plates will be fastened to the top quoins at the corners, and joined at the Mitchell posts. Fastening is accomplished by predrilling holes down through the plate beam into the corner quoins and into the tops of the posts. For additional tensile strength, a truss plate is used on the top of the wall to tie any two adjacent plate beams together. Such a continuous plate beam gives an excellent strong flat surface upon which to build a floor joist system.

INTERIOR WALL TIE-IN

Some house plans may call for two walls to join partway along one of their lengths. The technique is

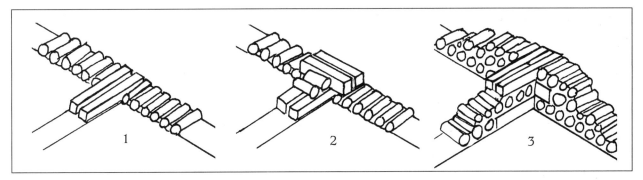

5-23. Joining one cordwood wall to another.

demonstrated in 5-23. Jack Henstridge says, "On a joining wall such as this, it isn't necessary to put in more than, say, three or four sections of tie blocks in an 8' wall."

Well, that's about it for built-up—or stackwall—corners. They are beautiful and strong, and they are the only way to use cordwood as load-supporting in a building with corners.

6
Cordwood Homes: Case Studies

Cordwood people are good people. We find this with the students who attend our classes and with the people who build their homes near and far. They are open people. They have to be open-minded in the first place just to consider building their house by this unorthodox method. They are not put off by friends, neighbors, relatives, and "professionals" telling them that cordwood won't work. They have the imagination to see right away the wonderful potential for an easy-to-build, low-cost, and fun way to build.

The owner-builders featured in this part share an almost childlike enthusiasm for their work. They'll tell anyone who comes along as much about their house-building experience as they can. This willingness to share probably stems from the memory of their own thirst for information at the time that they were first considering joining the ranks of wood masons. They are all glad to "spread the word" and add to the literature by sharing their experiences. Every one of them seems to learn some new technique or incorporate some new creative design feature. Easily the highlight of writing this book has been visiting people in their homes, and talking with other old friends on the phone, some of whom I hadn't spoken with in several years. I've learned a lot, and aim to share it with you in these pages.

In *Cordwood Masonry Houses*, I included 15 short case studies. Only three are repeated here: our own Log End Cottage and Log End Cave, and Jack Henstridge's landmark "Ship with Wings." Otherwise, the cast of characters is entirely new. Some are former students at Earthwood, some have built their

homes with little more guidance than the literature available at the time. A few of the wood masons tell their own story in their own words, adding a wonderfully personal touch.

POST-AND-BEAM EXAMPLES

1. Log End Cottage

As Log End Cottage and Log End Cave are often referred to throughout this book, I will limit their "histories" to the basic facts.

Jaki and I built the post-and-beam Cottage in 1975. Our previous building experience was the construction of a 12' by 16' shed in which we lived during the house construction, although I had worked as a mason's laborer for a few months in Scotland. Our log-ends are 9" white cedar, dried for one year before laying up. Most of the walls are built with the rather confusing mix of 12 sand, 1 Portland, 1½ masonry, 1 lime. This mix was not a success. Shrinkage and mortar joint cracks resulted. An experimental panel of 6 sand, 9 sawdust, 1 Portland, 1½ masonry, and 2 lime did not shrink. We have since simplified this mix and reduced the sawdust, as I've indicated elsewhere in the book. Complicated mixes invite error.

The house has about 500 sq ft of space in the basement, 500 sq ft on the first floor, and about 250 usable sq ft in the lofts. We spent about $5000 on

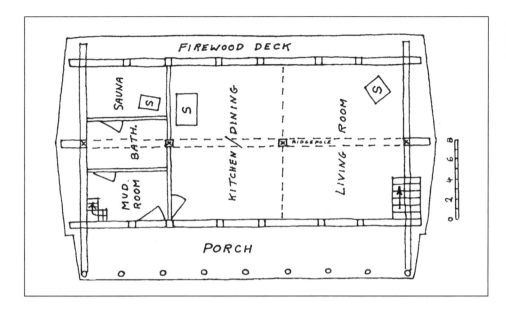

6-1. Log End Cottage, ground floor. Sleeping lofts over each end.

6-2. Log End Cottage.

6-3. Living Room at Log End Cottage.

materials and $1000 on contracting and labor. Half of the total went towards the basement and septic system, but even so, the house only cost $8 per sq ft of actual living space in 1975. We could not have done as well without scrounging and recycling a lot of materials. If I were to build this design again, I would use a floating slab instead of the basement, and I would build my lofts 2' higher to provide more usable space upstairs. Incidentally, the actual cost of

the cordwood infilling, including insulation, mortar and saw rental, was only $112, easily the most economical part of construction.

2. Log End Cave

Our second house, Log End Cave, featured 10" cedar cordwood infilling within the post-and-beam frame on the south wall and the north gable end. After 2½

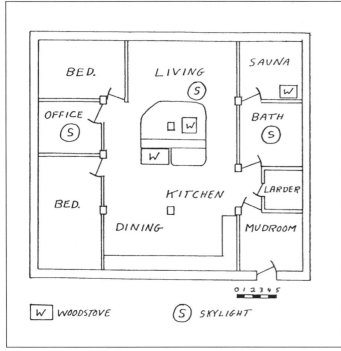

6-4. Log End Cave. (photo with floor plan)

years at Log End Cottage, Jaki and I did not relish the idea of living in a house without at least a few of these weird and wonderful cordwood panels.

The log-ends were force-dried in a sauna and next to the work stove during construction. Much of the cedar had been cut to 10' lengths and barked two or three years earlier. Wood shrinkage has been slight. The mortar at the Cave (6 sand, 8 sawdust, 3 masonry, 1 lime) has not shrunk, but is very roughly textured.

Log End Cave—Cost Analysis			
Heavy equipment contracting	$892.00	Styrofoam insulation	254.93
Concrete	873.68	Roofing cement	293.83
Surface bonding	349.32	Six-mil black polyethylene	64.20
Concrete blocks	514.26	Flashing	27.56
Cement	47.79	Skylights	361.13
Hemlock	345.00	Thermopane windows	322.50
Milling and planing	240.26	Interior doors and hardware	162.80
Barn beams	123.00	Tools, tool repair, tool rental	159.43
Other wood	167.56	Miscellaneous	210.31
Sheetrock	72.00		
Particleboard	182.20	Materials and contracting cost of	
Nails	62.88	house, landscaping and drainage	$6750.57
Sand and crushed stone	148.21	Labor	660.00
Topsoil	295.00	Cost of basic house	$7410.57
Hay, grass seed, fertilizer	43.50	Floor covering	
Plumbing parts	124.95	(carpets, vinyl, etc.)	309.89
Various drain pipes	166.23	Fixtures and appliances	507.00
Water pipe	61.59		
Metalbestos stovepipe	184.45	Total spending at Log End Cave	$8227.46

The only problem with the home was condensation which would appear at the base of the back walls during late spring and early summer. This was the result of warm moist internal air hitting the still cold base of the wall and floor near the vicinity of the footing. This is the "thermal lag," well known to students of earth-sheltered housing. Dew point would occur. This condition has never developed at Earthwood, where the footings and floor were wrapped with Dow Styrofoam®, not just the block walls as at the Cave.

Another change we would make would be to have a second means of ingress and egress. Even a chipmunk knows enough to put a second escape route into his bungalow.

The detailed cost analysis of our 910 sq ft house (1036 sq ft gross) is reprinted from my book *Underground Houses* (Sterling, 1979). Since the chart was completed, we spent about $350 more to finish off the place, or $7760 for the basic house. This works out to about $8.50 per usable sq ft of living space in 1976 dollars. About 2000 man-hours were required to build Log End Cave, 1500 of my own, 500 of others; some of it paid, some volunteered.

The house exceeded expectations during our three years there. It was light, bright, airy, easy to live in, and easy to heat. We burned about three full cords of wood per year, mostly in an oval cookstove, which also supplied our hot water. We used propane gas for refrigeration and summer cooking, and got our electricity from a Sencenbaugh 500-watt wind plant.

Presently, there is a couple buying the Log End Homestead on land contract. Desiring a larger house, they removed the earth roof, flattened the roof pitch, and added a two-storey conventional home on top. So Log End Cave does not exist as such any more. It has metamorphosed almost beyond recognition. But it has spawned many siblings throughout the country, from as close by as Champlain, New York, just 25 miles away, to as far away as Minnesota, Mississippi, and Alabama.

Year of main construction work: 1978.

3. Flatau's Plateau

Richard and Rebecca Flatau's beautiful home of red cedar log-ends in Merrill, Wisconsin, is one of the best known, thanks to articles in *The Mother Earth News* and *American Country*, and Richard's own little self-published book entitled *Cordwood Construction:*

A Log End View. The short quotes that follow are from these three sources.

During the "notorious winter of '79," Richard formulated a financial plan to build a mortgage-free home. After studying all sorts of alternative building methods, including log, rammed earth, stick frame, earth-sheltered, geodesic, and active and passive solar, Richard decided on cordwood masonry because it seemed to be the only method which would allow him to build within particular financial

6-5. (above) Flatau's Plateau, solar room.
6-6. (right) Inside the Flatau home. *(Photos credit, Richard Flatau.)*

parameters. His plan allowed $5000 to erect the shell and $15,000 for the finished home. The capital would come from savings, life insurance policies, paychecks, and equity from the sale of their mortgaged home.

The Flataus researched and planned... and planned and planned. Thirty revisions and several erasers later, they arrived at a 1064 sq ft (actual usable)* rectilinear post-and-beam plan with an 8 in 12 pitched roof sheltering an extra 560 sq ft of upstairs space built into the pre-engineered roof trusses.

The foundation was a thickened edge floating slab, poured by a local contractor. Anchor bolts were left sticking out of the concrete at the perimeter. Richard fastened two-by-twelve pressure-treated planks all the way around for a starter plate. This plate serves to keep the cordwood masonry off the slab—this strikes me as insurance against potential wood expansion problems—and makes it easy to fasten the many posts firmly to the foundation. The Flataus used posts—red cedar milled on two sides— about every eight feet around the perimeter of the 30' by 40' home. The tops of the posts were tied together with a doubled two-by-ten plate. Erecting the frame was, in Richard's words, "easier than we thought it'd be and demanded only the most basic carpentry tools and skills." It went up in "no time."

The cordwood masonry took somewhat longer. Richard advises would-be builders to consider the time factor carefully: "Stackwood construction is an inexpensive building method, but you'll be swapping hours for dollars. It took us over two months (after work and on weekends) just to put up the walls, not counting time spent gathering and cutting wood."

The Flataus bought their red cedar from four different sources. The 14 full cords averaged $25 apiece in 1979, with Richard hauling away the uncut wood himself. But it was already partially dry and almost totally barked, quite an advantage. A cost breakdown for the cordwood walls themselves—Richard kept accurate records—was $350 for wood, $140 for cement, $35 for 11 cubic yards of clean sand, and $80 for lime. The sawdust used as insulation and as a mortar admixture, was free. All told, the cordwood

walls cost $605, which is not bad for wall structure, insulation, thermal mass, interior finish and exterior finish, all at once.

The home was securely covered by mid-August of 1979, and all the windows were popped into their frames. Total cost so far was $4755, or $245 under Richard's financial plan. They worked on the house through the winter of 1979-80, but were stalled by a virtual standstill in the real estate market. Their home in town wouldn't sell, a grubstake they'd counted on to finish the home. Richard again: "Caught between a rock and a hard place, we chose to borrow $8000 from relatives—to be paid back upon the sale of our city house—rather than attempt to weather the economic storm in our old place while the new house sat in limbo."

The story ends on a happy note. After six months, they found a buyer for the city home, repaid their debts, and moved into the home in October of 1980, about a year and a half after breaking ground. The total cost was $14,955, $45 under the original plan. Very soon, however, the Flataus "splurged" and spent an additional $1020 for an improved water delivery system.

In 1983, Richard and Rebecca added a large solar room of 408 sq ft, also using cordwood masonry "panelling." The exterior of the solar room was simply cheap CDX plywood covered with chicken wire and leftover red cedar log-end disks about 4" thick. The disks were nailed to the plywood. Cordwood mortar filled the spaces between. The effect is to architecturally tie the solar room to the rest of the building. With a core of seven 34" by 76" "factory imperfect" thermalpane window panels ($16 apiece!), the resourceful owner-builders built the addition for just $1665, just $4 per sq ft, down from $10 on the main house. Later, they built four cordwood pole barns, getting their sq ft cost down to $1.21!

In a recent letter, Richard says, "We've met hundreds and hundreds of bona fide special souls who've come down to see our cordwood abode. Some have tried to buy it, others have been inspired to build, and others have lent ideas, thoughts, gifts, and messages to weave into the warp and woof of our lives."

---

* Note that a tax assessor measuring external dimensions would arrive at a figure of 1200 sq ft downstairs, but there is a 136-ft perimeter of 12" cordwood walls that reduces the downstairs to 1064 usable. With conventional framing, wall thickness doesn't come into the equation with quite so much impact. This is worth thinking about when you plan your house, and well worth pointing out to sympathetic assessors.

4. Harold Johnson

Harold Johnson, recently retired, lives with his wife Rita in Montreal. But they have a second home just over the border in Bellmont, New York, with 16" poplar log-ends laid up within a post-and-beam frame. Harold built the home on an old abandoned log house site. Most of the old home was beyond repair, but Harold was able to use many of the logs for posts, beams, and even some log-ends. Some of the framed portion of the old house was retained, sided in vinyl, and became a garage.

The two bedroom house is beautifully finished. Harold may safely be accused of being a perfectionist, and I don't think there would be any argument from Rita. The home is 28' by 40' outside dimensions (1120 sq ft gross, about 950 sq ft usable), took five years of part-time work on weekends and holidays, and together with the attached garage and a completely new basement, cost about US$25,000 to complete.

The resourceful and inventive builder came up with several innovations during construction:

(1) As owner of a pulp mill, Harold had a good sense of the nature and characteristics

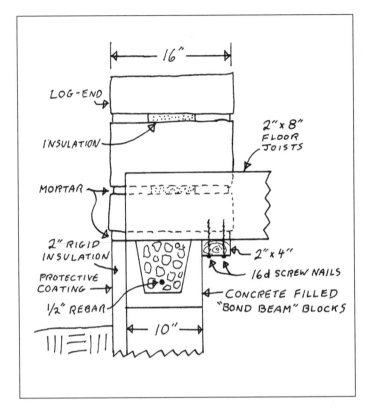

6-9. Harold's method of building a 16" cordwood wall on a 10" - wide surface-bonded block wall. The two-by-four plate is strongly fastened to the underside of the floor joists with screw nails. Log-ends are laid between joists and are partially supported by this plate.

6-7. Harold and Rita Johnson's house, Bellmont, New York. *See* page E of the color section.

6-8. White siding covers the old part of the house, now a garage.

of wood pulp. Instead of sawdust, he added slurried wood pulp as an admixture to his mortar. The mix was a success; the mortar is strong and shows no signs of shrinkage.

(2) The log-ends shrank somewhat, but Harold applied Perma Chink to the exterior mortar joints. He tried several different brushes, but found that the best was a fairly stiff-bristled painting-type brush about 2" wide and ⅜" thick. The application is a great success, closing wood shrinkage with a good-looking and flexible chinking. See 2-6. .

(3) Harold built his 16" thick cordwood walls on a surface-bonded block wall only 10" wide. While I do not normally recommend this, I must say that Harold's home obviously has no instability problems. His detailing is the key, shown in the following diagram on the opposite page.

(4) To brighten the interior, Harold painted a white mixture of joint compound and latex paint onto the cordwood walls. The cordwood texture of the wall still shows though the white coating, and the walls are very attractive. Siegfried Blum, of Priceville, Ontario, did the same sort of thing on the interior of his single-storey Earthwood house, but left a few prominent log-ends showing through the white background as highlights. These could be oiled with linseed to really accent the color.

5. Thomas and Anne Wright

Owner-built in 1986, this post-and-beam house in Rockmart, Georgia, has generated a lot of local interest. The 9½" pine log-ends look good over the stone-faced poured concrete basement. A few of the larger logs shrank, but Tom and Anne caulked around those with a mortar-colored caulking compound.

"Being employed by the railway, I used 12" by 16" bridge timbers for my corner posts," says Tom. Old boxcar flooring (two-by-eight tongue-and-groove) was used throughout the house for rafters,

6-10. (above) The Wright home, Rockmart, GA. *See* page F of the color section. 6-11. (right) The Wright home, Rockmart, GA. *(Anne Wright photos)*

ceilings, floors, doors, and the gable end planking. "We like the natural country look, easy upkeep, and economical heating and cooling. It just feels like home. Hard work, but worth it."

Thanks to Anne and Thomas Wright for the excellent pictures. They tell the story better than any further commentary from me.

6. Geoff Huggins and Louisa Poulin

We began construction of our cordwood underground house after having completed a small cordwood cabin as a warmup project. The cabin was first a weekend retreat from the city, then served as home while we built the house. The cabin is post-and-beam style, using native Virginia rough-cut oak, and the cordwood is six-inch-long Virginia pine. The relatively mild winter weather in our region does not call for a very large R-factor for walls.

The cabin is only 12' × 16', which made for cozy living for two during the house-building phase. Since we had no previous construction experience, completion of the small cabin proved a valuable boost to our confidence.

My engineering background allowed me to quickly absorb technical data as I researched cordwood and underground housing literature written by Rob Roy and others.

We had both lived in the Washington, D.C., area, working at good jobs for several years, so we were able to accumulate sufficient savings to build the house without obtaining a loan. Our budget required that we provide most of the labor. The only hired help we've had is the excavator, to dig the hole in the hillside, and a concrete finisher to lay the slab. Our financial outlay to construct the house was about $15,000, of which only $1500 was labor.

The major cost has been the large number of hours we have put into the project. It is difficult to estimate the cumulative time invested, as there has been the parallel task of creating and maintaining a homestead (clearing, gardening, orchard, vineyard, vehicle maintenance). Also, because we were far removed from our city friends and had made no local contacts yet, we were unable to create local work parties—we were on our own. Cordwood construction is a labor-intensive approach to house construction. That is its major cost, or drawback, in the eyes of some folks. Our choice was to hold dollar costs to a minimum, use locally procured materials as much as possible, and be lavish with our own time. Cordwood is an ideal way to go, with those guidelines.

We have been at it for seven years now (since the spring of 1984), so our pace is not the most rapid. Some of it is due to the additional homestead activi-

6-12. Geoff and Louisa's cordwood/underground house. *See* page E of the color section.

ties listed above. But equally responsible for what some would consider very slow progress is the feeling that "haste makes waste." We are building an unconventional house, trying to use environmentally friendly materials and be as compatible with our surroundings as possible. The nonstandard approach often means that there are no precedents for us to follow—we must come up with our own unique solutions. All of these factors create a slow building pace, one that gives full and careful attention to each step of the process. Sometimes this means that weeks go by without much of any progress, as a problem slowly simmers on the back burner.

We hope that this approach minimizes mistakes. Though we have been building for over seven years, we have been living in the incomplete house for five years. When will it be completed? We have long ago faced the issue that we are not even sure how to determine "complete." It became a non-issue when we adopted the Chinese maxim: "Man who finish house, die."

The house has 1000 sq ft of floor space—easily sufficient for two people. The only interior walls will be those enclosing the bathroom; the remainder of the house is open. This, with a 10½ ft ceiling in the center, gives the house a feeling of being larger than it is. The cordwood walls are post and beam, using tree trunks from our land as posts. These tree trunks dictated the wall thickness of 8 inches—a bit smaller than desired, even in Virginia's temperate climate. Because the core of the wall is insulated with sawdust, however, the walls have an overall R-factor of 8. While this is not very high, it is adequate when one considers heat loss through the walls compared to that through the windows. Even with double-pane windows, over four times as much heat is lost through the windows as through the walls in our house. (R is about 1.8 for double-pane windows.) Thus, a thicker wall would minimally reduce total heat loss, unless the windows were insulated at night.

As the underground walls have been backfilled bit by bit over the years, our energy requirements have dropped. Our sole winter heat source is wood, and the amount of wood burned has steadily fallen from 3½ cords our first winter to 1½ cords last winter, when the walls were backfilled only halfway up.

The cordwood used in the walls is Virginia pine, which is workable and relatively fast-drying. Its major disadvantage is its radial shrinkage rate, which

is twice that of white pine and half again that of eastern red cedar. Its shrinkage rate is nearly as large as red or white oak, both also plentiful in these parts.

Thus, although we were able to minimize mortar shrinkage by incorporating a good quantity of wet sawdust into the mix, we experienced some wood shrinkage. We were able to reduce wood shrinkage problems by adequately drying our logs (in somewhat less than a year, 8" logs dry quite well), and then treating them with oil, so they absorbed less water from the surrounding wet mortar. Our oil treatment is 50 percent used engine oil and 50 percent kerosene—the latter for thinning, to aid penetration. This treatment also gives the logs a brown color, so they don't fade to grey.

After the walls finally dried, we had a number of gaps to chink around the logs. Recognizing that few materials cling to both wood and mortar, and further that the logs would experience small dimensional changes from season to season, we sought a flexible, generally adhering chinking. We chose white silicone caulk, that we pushed into the gaps. While the caulk was still sticky, a soupy mortar mix (without wood chips) was painted on and forced into the exposed surface of the caulk, disguising the white caulk when it dried.

7. Jeff Corra

Jeff Corra had built nothing more ambitious than fences, counters, or workbenches prior to building his powerful "double-wide" Log End Cottage in the city of Parkersburg, West Virginia in 1983. The framework, seen clearly in 6-13 and 6-14, is made up of two eight-by-eight oak posts standing side by side every eight feet around the home. This method allows for the 16" thick walls necessary for northern climates. Jeff's log-ends were six months' dried oak. The larger pieces shrunk somewhat and were caulked. The cost of the home was about $14,000, about $2000 of which was for hired help. Again, costs of around $10 per sq ft were achieved.

Jeff's comments: "It's a great feeling to live in a house you've built with your own two hands out of materials from nature. The stone in the fireplace and chimney was collected from a local creek. Both gable ends of the house are all glass. I saved money for about ten years and collected materials for a year before beginning construction. The lack of any interest or loan payment helped."

6-13. Jeff Corra's "double-wide" post-and-beam frame.

6-14. Jeff's stone fireplace and split cordwood wall.

CURVED WALL EXAMPLES

8. Bob and Aggie Evans

Jaki and I met Bob and Aggie Evans just a few months before starting Earthwood. They had just moved to the North Country, had bought a piece of land in West Chazy, and were interested in building a round cordwood house, having seen some designs in *Cordwood Masonry Houses*. "We want to see a round house first," said Bob. Jaki and I had had the advantage of visiting Sam Felts' first house in Adel, Georgia, and felt good about round, but we decided to take a two-hour drive with Bob and Aggie up to Chesterville, Ontario, to visit a 40'-ft diameter cordwood house we'd heard about. We were all

impressed, and Bob asked me to help him design the structure of a 45' diameter home. They needed that extra space for their family. Increasing the diameter by 12.5 percent actually increases the area by 26.6 percent. This was important with three of their four kids still at home.

Their project started about the same time as ours, the spring of 1981. Bob owned his own business in Plattsburgh, Bob's Instant Plumbing. ("What is 'instant plumbing,' Bob?" "Easy," says Bob, "just add water.") Instant Bob decided on a floating slab foundation, and ran all of his plumbing and electrical conduit under the floor. He advises running water lines inside 4" sleeved "S & D" pipe. His electrical conduit pops up as needed inside internal walls.

The slab was poured professionally and monolithically, floor and footings at the same time. The contractor used two internal "fences," made of

6-15. "Aggie's House," West Chazy, New York. *See* page E of the color section.

staked two-by-fours on edge, to break up the pour into three manageable sections. The two outer pours were done first, the fences removed as the first pours stiffened, then the middle third was poured. As the diameter of the footing exceeds 40', an expansion joint should be included in the pour, best placed between the 4" floor slab and the 8" thick footings.

As Bob's Instant Plumbing was a new and growing business, Instant Bob had very little time to spend on building. Before leaving for work each day, Bob would plumb up—who better to do this than a plumber?—a curved form which we had designed together. The form was made of 1" by 10" base and top plates, with two-by-four studs following the curve of the building, 16" on center. (*See* 12-1.) That was the section which Aggie would build each day, with 14-year-old son Jeff mixing the mortar. A young grandmother, Aggie laid up virtually every log-end in the home. The one drawback to a form such as this is that the studs make for difficult pointing on the outside of the wall. Aggie did her best with just her rubber gloves. Upon close examination, it is possible to see the fossil remains of the framework on the building's exterior. All in all, though, it has to be said

that the use of the forms was a success, and greatly speeded up a slow process, by eliminating the need for constant checking with a level.

The Evans family moved into the home in late October of 1981, about six months after the slab was poured. The home was 1590 sq ft gross (1407 sq ft within the cordwood walls) and cost about $14,000, which included labor payments for the slab, the central brick mass ($1600) and installing the heavy roof ($2100). Again, the almost unbelievably magic number of $10 per sq ft was realized. The central brick mass was supposed to have been round, but they couldn't find a mason with enough imagination to build round, so it is square.

Several years later, Bob built the trapezoidal shaped addition that shows in 6-15. This addition follows the radial lines of the heavy five-by-ten rafters of the main house, which is much more pleasing to the eye than when a square building is added to a round one. The roof tie-in is also much easier. Round buildings are not easy to add on to, so either make it large enough the first time, or plan for an architecturally harmonious addition. Consider the roof line tie-in very carefully. Go radial.

9. Kim Hildreth

A clergyman friend living near Jay, New York, about 35 miles south of us, and visiting Earthwood for the first time, surprised us by saying that a nearly identical house was built near his home. "Are you sure?" we asked. It seemed most unlikely that such a house could be built so close to us in New York's sparsely populated North Country without our knowing about it. "Oh, yes, quite certain. And a beautiful home it is, too. Belongs to Kim Hildreth."

It wasn't long before I called Kim and arranged to visit, and Jaki and I showed up at the home a few days later. We learned from Kim that he'd built the home with only my book *Earthwood* as a guide. Much of that book remains in Part Two of this one. The major differences were that Kim eliminated the earth roof in favor of a steeper-pitched shingle roof, and made only a half storey upstairs, which he calls "the loft." Also, the house is not earth-bermed.

Kim, a professional landscaper, did meticulous work inside and out of the home, not just with the cordwood masonry, but with his carpentry, stone masonry, everything. The home has been featured with beautiful color pictures in *Adirondack Life* magazine.

I revisited the home recently to take the pictures shown in this article, but did not meet up with Kim. Later, we spoke about the house on the phone. During my visit, I noticed some rune-like inscriptions in the mortar joint up near the eaves. "9-19-82" said one. Others froze in time the names of people who must have been there helping or at least standing by cheering on that date: Kim, Ona, Julia, Trevor, Tim, Chuck, Rich. When I mentioned my observation to Kim, he laughed. "Yeah, that was quite a time. We shingled the roof three months later, and the house was protected for the winter. It was only a shell, though, with a lot of finish work to do inside. As we were living comfortably on site in a mobile home, we took our time to get things right. We moved in during the fall of '83."

I learned some cordwood construction details in our phone conversation. The first course of wood is northern white cedar; all the rest of the log-ends are red pine. The cedar on the first course, particularly with a round house built on a slab, may give some protection against wood expansion problems later on. The expansion problem is most likely to occur

6-16. (above) Hildreth house, south elevation. *See* pages D and G in the color section. 6-17. (right) Window and cordwood detail at Hildreth's.

by water collecting against the first course of wood, and white cedar is one of the most stable of species. Some of the red pine log-ends have beautiful five-sided star patterns as a natural feature. This design occurs when a log is sliced right where five small branches come together, Another of Kim's comments on red pine is that it does not split straight, so split log-ends tend to take a twist through their length, making them hard to use. For this reason, Kim used mostly rounds in the wall. His mortar mix was the same that we used at Earthwood and recommend in this book. I observed no mortar shrinkage cracks in the wall. The mortar cavity was insulated with loose cellulose.

Kim uses five 16" face cords of wood to heat the home each year, all of it passing through a Shenandoah airtight woodstove. Copper tubes, woven around the round central stone mass, preheat

the domestic water very well, which saves on the cost of water heating. The floor is over 5" thick of concrete, with a slate surface on top, and 2" of Dow Styrofoam® beneath. From 10 A.M. to 2 P.M. on sunny winter days, the floor is "charged" by solar energy pouring through the large south-facing windows. There is a floor-to-peak duct system by which a fan can pull hot air out of the high point in the loft and exhaust it out onto the slate floor downstairs.

10. Earthwood, Jr. by Paul Mikalauskas

I decided to build my own earth-sheltered house while reading Rob's book *Underground Houses*. It just seemed so sensible. With interest rates rising, along with everything else, the only way I could afford my

6-18. (above) "Earthwood, Jr." *(Paul Mikalauskas photo.)*

6-19. (right) "Earthwood, Jr.," interior. *See* page F of the color section. *(Christine Hobart photo.)*

own house in 1981 was to build it myself (with a little help from my friends.)

I digested books on underground houses and owner-building. After two pilgramages to Earthwood seminars, I decided to build a one-storey version of the Roys' own home.

After a year of searching, I found an ideal five-acre lot in central New Hampshire. I kept my job in Massachusetts for two more years, drawing plans at night and commuting 120 miles to the site on weekends. Friends Dave Larson, John Ryan, and I spent many weekends carving a house site out of dense woodlot. Each white pine would get peeled of its bark, split if necessary, and stacked and covered to dry.

One of our first projects was to build a live-in shed and tool store, enabling us to work the entire weekend without having to drive home each day.

During the next year, I had a well drilled, built a septic system, and poured the floor slab. I found a local job and moved into the shed, in order to get more work done.

Living in the shed was tough. I hauled water from the well in milk jugs, used an outhouse, and kept the wood stove going constantly to keep my dog from freezing. People warned me that as soon as I moved into my permanent home, work would screech to a halt, so I wanted to make sure it was finished first.

In the spring we started the surface-bonded curved block wall, just as Rob describes it later in this book, but this took much longer than it should have, because the blocks were wildly inconsistent in height.

The race was on. Summer had come, but it rained every weekend and we knew we had to keep moving to finish the shell before winter.

By the fall of 1985, we had finished the below grade portion of the wall. The surface-bonded blocks looked real good, and we were inspired. The cordwood masonry and the 5'-diameter chimney went up in '86. I'd built the door and window frames ahead of time and spent my spare time during the week setting up for weekend work parties. It is important to have all materials close by, but not in the way. John and Dave would come up on Saturdays, and work would fly. My parents also joined us frequently, and their support was very important to the ultimate success of the project.

The only work I hired out was the chimney.

Although I can do masonry, the chimney was going to serve as the central pillar, so I wanted it strong and good-looking. I hired my friend Jack Wood for the job, and kept him supplied with mortar, stones and flue. We used attractive stones for the outer surface, filling in with rubble. A week's work resulted in a beautiful and functional focus to the center of the house.

We got quite efficient at cordwood masonry. I'd mix mud, John would select or split the right log-end, and Dave would lay down mortar and insulation. Our manageable panel size got bigger and bigger, but we were careful not to wait too long before pointing. Soon we'd installed the window and door frames, lintels and plates, and were ready to start the octagonal post-and-beam inner framework.

Posts were cut from 12"-diameter white pine that we'd cut while clearing. The other timbers were hemlock, cut at a local sawmill. I had the five-by-tens and ten-by-tens on site, and covered so that they would be somewhat dry when we needed them. They were still incredibly heavy, but a borrowed hand-operated forklift made those timbers fly.

With the help of a big work crew, we erected the entire roof structure in one day. The following day, we decked the roof with tongue-and-groove two-by-sixes. The first snow came the next night.

We applied the Bituthene® waterproofing membrane very close to its 40° F. design limit, which made me nervous. But I couldn't wait. Winter was coming fast. It snowed again that night.

I had plenty of work inside, though. I connected a stove to the chimney and started building interior partitions.

I panicked after discovering a leak in the roof. Evidently, snow was melting on the uninsulated roof and damming at the unheated overhang. Water was working in under the Bituthene®. We had to scrape the snow and ice off the roof, trying not to harm the membrane. We repaired the leaks with lots of mastic and then installed 4" of Styrofoam® insulation. We covered the insulation with 6-mil black plastic. The earth would have to wait until spring.

Work continued inside for another year. Interior walls went up and closets were built. Kitchen cabinets were made from scrap lumber. After sanding and a hand-rubbed oil finish, they looked great.

We covered the entire floor area with unglazed Italian tile, which I'd obtained at a great bargain. We even had enough left to surface the kitchen counters and the breakfast bar. The tiles took over a month, but I'm pleased with the results.

We hid most of the wiring in interior walls and in a channel which runs along the top of the exterior cordwood wall, between the plates. In some places, we installed inexpensive EMT tubing and surface-mounted boxes and painted them to look good.

Setting a date for the housewarming party lit a fire under my trousers. The night before the party, Dave and I were up until 2 A.M. installing the submersible pump in the well. I was still connecting the kitchen sink as guests arrived. I spent that night, my first, in my new house.

Building my own house gave me an overwhelming sense of accomplishment. The place fits me like an old shoe, and it is comfortable and efficient. Some heat comes from the sun, but most comes from 2½ cords of firewood burned each year in a catalytic wood stove. In hot weather, the earth berm on the north side and the earth roof keep the inside cool.

My earth-sheltered home is environmentally sound, blends in with the site, and ages gracefully. It is affordable. Taxes and mortgage combined are less than apartment rentals in the area. The mortgage should be paid in five more years.

My other debts will be paid off when John and Dave build their own homes.

11. Mushwood

With a name like "Mushwood," it has to be good ... with apologies to Smucker's. We call our summer camp at Chateaugay Lake *Mushwood* because of its shape—it looks like a *mushroom* and because the first floor is *cordwood*. The second floor is one of George Barber's geodesic domes.

The downstairs part of the building has an outside diameter of 22'. Subtracting for the 12" cordwood walls, there are 314 sq ft usable downstairs, which space is devoted to two small bedrooms, a bathroom (having a shower, not a tub), some storage space, and circular stairs to get you up into the dome. The upstairs space is open-plan—living, kitchen and dining—and actually has about 615 sq ft usable. Increasing the inside diameter from 20' to 28', then, almost doubles the floor space. (It is more than dou-

6-20. "Mushwood," Chateaugay Lake, New York.

bled if the "unusable" floor space of the open stairway is subtracted.) This is the miraculous function of the radius being squared in the pi · r² formula for a circle's area. So the top floor is airy and spacious, with a full 14'-high ceiling, while the downstairs space is privately functional. We think it makes a great plan for a lakeside cottage.

Construction was a little out of the ordinary, even for cordwood masonry. First, we cast a monolithic floating slab, 22' in diameter. Next, George Barber helped us to put up one of his 29'-diameter domes, covering the entire work site. Six-mil greenhouse plastic kept the rain out while we proceeded with the cordwood work, much of it done during workshops. The plastic skin took us through the first winter.

We set up a 26"-diameter pine tree trunk in the center of the building as a post for the 16 floor joists to rest upon. This tremendous post, given to us by friend, neighbor, and logger Pat Collins, is quite a feature of the house, its colors and textures set off by three coats of polyurethane. When we got the wall up to floor-joist height, we set our plates as discussed in Chapter Four, and installed the joists. To accommodate the very long (18') rafters, we had to cut holes in the greenhouse plastic, and for a while the building looked like something from outer space. But we were able to finish the cordwood "snowblocking" (the space between joists) while still under cover. This snowblock work is much easier to do before the floor planking is installed.

Then we dismantled the dome, and with a big push, planked the eight flooring facets of the building, and reerected the dome at the second level, where it remains as the upstairs framework. A quick sheathing of plywood and 15-lb felt took us through the second winter.

A sitting deck overlooks the lake, accessible from the upstairs by sliding glass doors. An outside stair from the deck provides the second means of egress and a quick route to the lake. We built the front deck around a large cherry tree, a hole in the deck provided for each of its two trunks.

There is also a continuous 3'-wide deck all around the dome at the second level, kind of like the widow's walk on the homes of Nantucket whaling captains. This is not intended as some fancy and expensive architectural feature, although it seems to add positively to the lines of the building. Its main purpose is functional: one needs a place to stand in order to build the dome. A very useful side benefit of

the walkaround deck is that we can always find a sunny (or shady) place somewhere around the circle to relax.

The Mushwood design has been one of our most popular. It is an unusual but esthetically pleasing home. We even built one for a client from Brooklyn, on a wooded lot about a mile from Earthwood. We like the open spaciness of a dome. And they are easy to heat with a wood stove placed in the center. Our dome walls have three inches of Styrofoam® in them, only about R16, but the relationship of skin area to space enclosed is optimum in a dome, and rising hot air hits the center of the ceiling and then flows down the walls in a natural convection current. For a summer cottage, and occasional winter use, the building is perfect.

The main drawback is that although the dome framework goes up in a day, and the 40 sheets of plywood needed to cover it can be installed in another day, the building takes forever to finish. We've been six years working on ours, on and off, and it still isn't finished as I write, although it should be by the time of publication. Some friends of ours, who built a home composed of two domes, a connecting "in-between," and a kitchen addition, took four years to build their home. Domes take time. Cordwood takes time. Combine the two, and... well, it's got to be a labor of love, or it will turn out the other way around. Are we glad we did it? Yes. Would we do it again? No.

Cost-wise, even with electrical wiring (professionally done), flooring, cabinets (recycled) and all plumbing (but not the well and septic system), we'll finish the place for under $10,000, or less than $11 per sq ft, ,not bad in today's world.

12. The Ship with Wings

Jack and Helen Henstridge's 150-year-old frame house burned down in 1973. Compounding their problems was the fact that Jack had just lost a good job as a company pilot because of a business recession. They were truly up against it. Their assets were: a good piece of land on New Brunswick's St. John's River in Canada, $8000 in fire insurance money, and plenty of kids and friends to help with building a house. The family moved into an old schoolhouse and Jack began to plan his new home. The end result was *The Ship with Wings*, probably the best known of all cordwood homes, having appeared in several

6-22. "The Ship with Wings," plan.

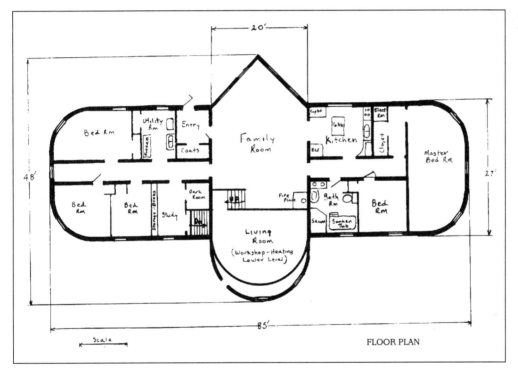

FLOOR PLAN

American and Canadian periodicals, including *The Mother Earth News*.

The cordwood was about 80 percent softwood and 20 percent hardwood. Jack was in real need of shelter and wasn't too fussy about the kind of wood he used. The 9" log-ends were laid up green, and considerable shrinkage resulted. Jack spent about $6000 on materials for his house and another $500 for electrical contracting, which comes to $2.50 per sq ft. The east wing was never completely finished and is used mostly for storage.

Jack describes his home as "an experiment that worked exceedingly well and proves that it is quite possible to build with green wood and a 9" wall. I couldn't have spent much less unless I used a clay mortar in the walls. The house is much larger than necessary, but the method is proved out. Green wood requires more 'afterwork' and the 9" wall, while adequate, is just too hard to stack up. Sixteen-inch walls would be ideal."

All the trials and tribulations of building the Ship with Wings are detailed in Jack's entertaining book, *Building the Cordwood Home*, listed in the bibliography.

Had the late Canadian balladeer Robert Service met the venerable Henstridge, he might have been inspired to write:

The Ballad of Cordwood Jack

It was twenty below that fateful night the Henstridge house burned down,

The flames you could see a mile away, lighting Upper Gagetown.

"I felt like a fool, but went back to school for a place to live," says Jack,

"The insurance was paid, but it wasn't much, and I started to plan my shack."

The plan was for logs, but through swamp and bogs, he couldn't haul 'em out.

"We'll hafta cut 'em short," said Jack. His wife then started to shout.

"You can't do that! They'll be too small!" she earnestly reported.

"We'll lay 'em crosswise," said Jack. "We'll stack 'em all up mortared!"

13. "Arkwood," by Sam Davis

I first came across cordwood building in a *Mother Earth News* article in the '70's. Little did I know that a seed was planted that would sprout into a round cordwood house in the Ozarks 15 years later. My wife Sue and I bought 35 acres of woodland on a river near Eureka Springs, Arkansas in 1982, and moved into a mobile home on the land in anticipation of building our own home. Cedar was plentiful and there was sand in the riverbank, so cordwood seemed the logical choice. I attended one of Rob's workshops in east Texas in May of 1986 and began the construction of "Arkwood" the same year.

We borrowed many of Earthwood's basic design concepts and added a few of our own ideas, such as building on a north slope to better meet our needs for summertime cooling here in Arkansas. The upper storey includes a greenhouse and solarium facing south. We used a post-and-beam frame on the upper storey walls, and filled in with 14" log-ends. We used an octagonal hip roof system with composition shingles instead of an earth roof. The ceiling ris-

6-23. "Arkwood," a round house with a post-and-beam frame. *(Sam Davis photo)*

es to 15', where it rests on a circular stone column à la Earthwood. At 2400 sq ft, and with two bathrooms, the cost was about $25 per sq ft.

We shared experiences well known to other owner-builders: the pride of building with one's own hands, and the satisfaction of accomplishing a long-held dream. There were also the pitfalls of money problems, the endless toil, the frustration and strain on family relationships. To be honest, this project was not a bed of roses for us. Anyone considering a project of this scale should definitely do a serious marriage–reality check. A tremendous amount of sacrifice and focus is required. The old adage, "The greater the sacrifice, the greater the reward" certainly applies here. In our case, I developed a severe form of rheumatoid arthritis in the middle of construction. (The disease was not building-related.) While this caused a year's hiatus in the project, and tested our resolve as a family, it somehow made us even stronger and more determined to meet the challenge. The building was aided by family (ages 3 to 83), friends, neighbors, and even passersby, if they lingered too long on site.

Cordwood building was well suited to our situation. We used materials from our own land, financed the project as we went along (4½ years until we moved in) and now have a debt-free, energy-efficient, comfortable, and—as visitors remark—"interesting" home. We've marked our page in the history of building, as this house should stand the test of time. As a neighbor said, "Looks like it could take a direct hit."

14. Earthwood West, by I.R. Lafon.

This is a true story. I'll let I. R. tell it in his own words:

In 1985, having made the decision to leave the stress of Silicon Valley, we acquired Rob Roy's *Earthwood* book. Studying *Earthwood* and several other building techniques confirmed our preference for Rob's plan. We purchased our property in Washington that summer, felled trees, and cut, split, and stacked the western cedar and Douglas fir logends to dry. The excavation was bulldozed and temporary power installed. We readied a shed site for a Cordwood Masonry Seminar which Rob conducted at our place later in the year.

To meet local seismic code requirements—we

6-24. Earthwood West in March of 1990. The 16-side post-and-beam frame was required to meet earthquake codes. *(I.R. Lafon photo)*

are in a seismic "zone three"—the following changes to Rob's plan were made:

- more than double the reinforcing bar (rebar) in footings,
- rebar reinforcing for the underground block wall,
- extra rebar in the buttresses,
- extending buttresses upward to support the roof frame,
- attaching buttresses to the wood framing with steel straps,
- adding post-and-beam framing to the cordwood walls,
- adding framing around the chimney, unloading the masonry.

We have made other minor changes as a matter of personal taste, such as moving the hot tub into the greenhouse.

Except for the concrete pumping and finishing, and the first storey of the chimney, we did all the work ourselves. One day, we had some friends help us lift the rafters for the price of a meal and some beer.

We spent a great deal more time on our version of the house than did Rob and Jaki, due mainly to additional code requirements and our own personal choices. There were times when it seemed that the end of construction would never come, but by late 1990 we were able to keep the house warm enough to live in, a great luxury compared to the tin can (trailer) we had been living in.

We used the money saved on labor to buy the highest quality components, such as windows, doors, and plumbing and electrical hardware. We spread our expenses across the time spent on the various tasks, a big economic advantage over paying to have someone else do the work. As Rob says, "Money saved is a lot more valuable than money earned, because we have to earn so darned much of it to save so precious little."

The cost of the house as it is now comes to $47,200—I keep track—or just under $24 per sq ft of living space, about half the fair market rate of $47.20 in 1985 when we started. Our lack of previous building experience was more than offset by our technical backgrounds and stubborness. It can be done! And we have even managed to retain our sense of humor, although our sanity suffered occasional spasms. And, yes, we'd do it again.

STACKWALL EXAMPLES

15. Cliff Shockey's Second House

For 15 years, Cliff Shockey, building in the cold plains of Saskatchewan, has employed his double-wide stackwall system, actually two cordwood walls

6-25. Cliff Shockey's second house. *See* pages A and H of the color section *(Cliff Shockey photo.)*

6-26. Stackwall corner detail at Shockey's. *(Cliff Shockey photo.)*

6-27. Shockey II interior. Note fine balance to cordwood masonry. *(Cliff Shockey photo.)*

separated by a 10" thick layer of fibreglass batt insulation. There is a plastic vapor barrier between the insulation and the inner (6" thick) cordwood wall. The outer wall is 8" thick. Both cordwood walls are laid up without an additional insulation cavity and Cliff uses the simplest mortar mix I have encountered: 3 sand and 1 masonry cement. He says he has had "very little" mortar shrinkage. ("I laid the wood in the fall and spring.")

Cliff, now retired to his life as a "hobby farmer," has built two houses and an office using this method, and if his roofs employ insulation in scale with his walls, one would assume that the buildings could be heated with love alone. "The double cordwood wall is the only way to go in this climate," says Cliff.

The cordwood for all the buildings was free: Cliff was given six miles of abandoned cedar telephone poles. All he had to do was remove them. The bottoms of the poles were creosoted and unacceptable as interior log-ends, so Cliff cut them into roofing shingles. The rest of the pole was untreated and made excellent dry, beautiful cedar log-ends. Cliff's skill and sense of artistic balance can be seen in the pictures.

Cliff's second house was completed in 1981, with an addition built in 1990. The basic home—somewhere between 1500 and 2000 sq ft, depending on whether you consider the 24" wall thickness—cost about CAN$43,000. ("I went more elaborate on the inside than necessary.") Cliff's masonry work is hard to beat, but with a wife named Jackie, how can he go wrong?

16. Sam Felts and Kevin Rencarge

School teacher Sam Felts built his 40'-diameter round house of red cedar in 1979. The home in Adel, Georgia (just 40 miles from the Florida line) was featured in *Cordwood Masonry Houses*. His adopted son, Kevin Rencarge, assisted during that project. Now retired, Sam has kept an active interest in cordwood. I'll let him tell the story of Kevin's home:

Shortly after being married, Kevin and Loretta began to dream of a cordwood home of their own. Their dream was of two storeys, a living room with a high ceiling, and a balcony looking down to a fireplace. I was appointed chief designer.

It was decided to use 16" cypress log-ends for the exterior walls, providing for four inches of mortar on each side and an insulation cavity for eight inches of sawdust insulation. Both the round style and built-up corners would be incorporated into the structure.

The front of the house is a two-storey semicircle, 24' in diameter. Half of the semicircle is a living room with a fireplace, the only source of heat for the house. The ceiling is 18' high, with a balcony overlooking the room. Wood blocks the size of bricks form the rear wall. Spiral stairs to the second storey wind gracefully around the polished trunk of a cedar tree, the hand rail formed from a wild grape vine. Two bedrooms, one over the other, make up the other side of the semicircle.

The "back" of the house is a stackwall-cornered

rectangle 16' by 36', and consists of a combination kitchen-dining area, a large utility room with outside entry and firewood store, a bathroom, and a large master bedroom with exposed beams.

The entire roof is surrounded by battlements with slits (embrasures) as a design feature. Arched entrances and windows also contribute to a feudal castle motif. An overhanging balcony provides a cozy and romantic spot to view the sunset. This wooden castle, with its 2000 sq ft of floor space, was finished in 1986 for a total materials cost of $20,000. Outside labor added another $2500, for a total outlay of just $11 per sq ft. Kevin and Loretta took a bank loan to build the house.

6-28. Sam's first house, built in 1979. *(Sam Felts photo.)*

6-29. Kevin and Loretta's Cordwood Castle. *(Sam Felts photo.)*

6-30. Child's bedroom in the Cordwood Castle has a wood-block wall. *(Sam Felts photo.)*

17. Irene Boire

(Author's note: Irene's story is more personal than technical. It is the story of an appreciation of simple values and of overcoming the odds. It is a story best told by Irene herself.)

A house is not just a stack of lumber. A house is a whole collection of memories put together with sweat and dignity. Downstairs, in the north wall masonry, are the fingerprints of children who lent a hand. Panels of cordwood show the artistic taste of many different helpers. Tiny pebbles pressed into the pointing and tools cemented into place as permanent souvenirs all sing a silent song of proud memories.

Elizabeth Mangle and I were college roommates. We shared a mutual desire to produce our own food and live simply. It occurred to us that the way to establish the life-style we had in mind was to simply haul off and do it.

I designed the house according to how we wanted to live and according to the ambiance desired. My first thoughts were of the heavy timbers and the many windows. I then planned a structure that would support the beams and windows. I thought about things like heat, storage space, and where to put the furniture. Elizabeth doesn't care much for drawing house plans so her design contributions were verbal in nature. A great deal of time was spent drawing floor plans and then cutting little cardboard furniture to scale. This cardboard cutout approach to house planning proved highly educational and gave us the assurance that the house would have the room we needed. Armed with a realistic and detailed plan for a structure, we looked and found a parcel of land. At some distance from the asphalt there was an old foundation, half fallen in and grown over. The road to the site had seen little use in forty years. The great advantage of having bought this remote site is that we now have the last house before the "wilderness."

The top of the old stone foundation exhibited what could be described as a disturbing degree of unevenness. I found cordwood masonry to be the ideal method for continuing upward. The plastic nature of "mud" allows for a custom fit. The west wall of the basement is an earth-bermed stone wall with twelve to twenty-seven inches of cordwood masonry on top. The south wall is only about one-quarter stonework, and that slopes radically; therefore, the remainder of the wall is cordwood. Both east and north walls of the downstairs are entirely cordwood. The cordwood cedar walls are twelve inches thick.

The upstairs is framed in rough two-by-sixes. The exterior walls are covered with Adirondack siding (wood milled on only three sides) and the interior walls are horizontal V-joint tongue-and-groove. Originally I had envisioned two stories of cordwood masonry because it is inexpensive. However, when I

6-31. Irene and Elizabeth's house. *See* page D of the color section.

6-32. Round window detail.

6-33. Irene's stackwall corner detail

was offered lumber at a very good price I capitalized on the opportunity. I had the privilege of going into the woods with our lumberjack friends and skidding out some of the trees which later became heavy ceiling beams. When the sawyers were rather short of time we offered to lend a hand and so we planed and shaped our wall boards.

While my vision of heavy timber rafters was a beautiful dream it was also a perplexing conundrum. With ridge pole construction I could lift and position my rafters one at a time, an important consideration when a crane is out of the question. But, exactly what holds them up there year after year? In general, the peak of the roof wants to fall down and the walls of the building want to fall out. Where the tie beams meet the rafters I designed a joint which, when chiselled out and fit together properly, not only resists these natural tendencies, but locks the heavy timbers together all the more solidly as they try to shift downward due to gravity.

Having solved the rafter riddle, how does water get into the house? Well, that is simplicity itself. One carries it in by the bucketful. How then does dirty dishwater get out of the house? One simply opens the door and flings it out. This system saves a great deal of money in pipes, fixtures, water pump, etc.,

but I must admit it can grow fatiguing.

The house is heated primarily by wood with some propane heat available as backup. The main heating unit is the masonry stove.

After taking such pains to create a heating system our electrical system seemed to fall into place quite easily. We produce our electricity by means of photovoltaic panels. While I know of individuals who go to great lengths to put their p.v. panels at exactly the perfect angle to the sun, ours are simply bolted flat to the south wall of the house. Here, they are safe from strong winds, falling branches, and snow cover.

Light and air are important to me, and the basement has so many windows that even on a dreary day one can still work comfortably without having to use electric lights. Upstairs the clear gable ends give a great deal of light and the illusion of great size and space. I wanted my house to connect me to my French-Canadian heritage—to have a sense of history about it—and so I designed the upstairs windows to reflect the style and craftsmanship of the windows one sees in Old Montreal.

While building the east wall, Elizabeth decided that a round window might be fun. I made a design and drew patterns on cardboard for the different components. Inspiration struck! Since my older brother had a variety of power tools, he could readily cut out all these components. So, I gave him the patterns and explained how many of each I needed. I expected he would deliver a bushel basket full of pieces but my brilliant big brother put the window together... and not with nails but with pegs! At a later date we installed Plexiglas cut round by the gentleman at the shop. Who says great works of art are not made by a committee?

18. Betsy Pugh

In the town of Ausable, New York, about 25 miles away from Earthwood, Betsy Pugh and her family and friends built their 27' by 37' stackwall chalet over a three year period. The first year's work consisted of cleaning up the site where an 1820 framed house had burned down, and making the old stone foundation ready for a new home.

The second year saw the cordwood walls get up about halfway, where they stayed covered through the long North Country winter. The third year saw the house completed, including one weekend roof-raising party during which the planks and black paper were installed over the whole building, and the shingles on one side. The home was occupied on Halloween, 1983. It has a full basement, a half loft and, like most cordwood homes, no mortgage. Betsy reckons the house cost about $20,000, which includes "materials, tools, and beer." This works out to less than $17 per sq ft, not counting the basement space.

Betsy featured two different kinds of stackwall corners which are a little out of the ordinary. The east gable end has corners made of nailed-together recycled two-by-eights, creosoted on the outside (see 6-35). The two western corners are made from doubled six-by-six cabin logs left over from someone's log project. At about five feet, these are definitely the longest quoins that I have seen. These, too, are creosoted on the outside only, and stained on the inside for a contrasting effect to the log-ends.

The walls themselves are 24" thick and include all kinds of fully barked wood, including pine, cedar, white birch, oak, and elm, 15 full cords in all. There is a lot of relief in the walls due to log-end length variations. Betsy says she would be more careful if doing it again, so that there would be less relief, and the inside walls would be easier to clean. Another word of advice to others is to use "sill seal" roll insulation between the wooden sill plates and the foundation walls. "A lot of draft comes in there," she says.

6-34. Betsy Pugh's stackwall house. *See* page G of the color section.

6-35. Betsy nailed this corner together.

7
Special Effects

There are many good reasons to build with log-ends: wood unsuitable for other styles of building can be used; the work is not as physically demanding as for a "traditional" log cabin; the wide wall makes for excellent insulation and thermal mass; and the total cost of the house is low.

Still, one of the primary appeals of cordwood masonry is visual; somehow, this building style is faintly reminiscent of the gingerbread cottages in storybooks. Together with a sturdy post-and-beam framework, the look is similar to that of the English "black-and-white" houses, and, at the same time, has a certain Scandinavian flavor. The log-ends have such a natural beauty of their own that it would almost take a conscious effort to lay up a wall that is not pleasing to the eye. The different-colored patterns of the end grain and the random checks that form during drying almost guarantee success. Almost.

But if time is taken, and attention to detail, improvisation, and, above all, imagination are exercised, your walls will be personal, unique, and very beautiful. I will present a few ideas in this chapter which I hope will illustrate that cordwood masonry can be a highly creative medium.

RELIEF

Recessed pointing by itself should ensure a three-dimensional look to the wall, but further interest is added by deliberately recessing a log-end or allowing one to protrude, especially if the log-end has some special color, texture, or natural design; or a log-end

can be wood-burned with a hot wire and featured in relief. A good example of relief is shown in 7-1.

VARIETY OF STYLE

When building Log End Cottage, we were able to obtain a couple of loads of scraps left over from cedar logs that had been used for traditional log cabins. The scraps ranged in length from 1' to 3' and were all uniformly milled to 6". For variety, and to make use of this valuable windfall, we built three

7-1. A good example of relief.

7-2. Two different styles of cordwood masonry.

12" to 14". These we laid together in a course to make a boot shelf for our mudroom.

PATTERN

We stayed with the random style throughout Log End Cottage, except for a few stained *beam-ends* that we placed in some of the square panels as focal points (7-3). Another idea is to take a few log-ends of the same size and to feature them in a design. The drawings (7-4) illustrate some of the possibilities, which, except for the pentangle, take advantage of the natural hexagonal configuration which log-ends of the same size will assume. To accent the design, log-ends 2" longer than usual should be used and the whole design should be allowed to extend 1" proud of the rest of the wall, inside and out. The center log-end could be regular-length, or it could be a "bottle-end."

Design work of this kind takes advance planning—special log-ends should be put aside for the purpose—and it takes time, but it's worth it. Some of the best designs in the house, though, will be those that occur naturally in the random style. We have one panel, for example, where small log-ends seem to cascade over a precipice and smash on a large log-end below.

panels with the log-ends laid up horizontally in courses. The result is shown in 7-2. By coincidence, the roof slope of one of the cabins was the same as that of our diagonals, and we only had to cut three or four pieces to fit. Some of the pieces were very wide,

7-3. Planning the placement of a "beam-end." Stained beam-ends add interest to rectangular panels. They should be centered in the panel and leveled. Centering takes planning and measuring. Ends A, B, and C were used specifically to bring the work up to the right height for the beam-end.

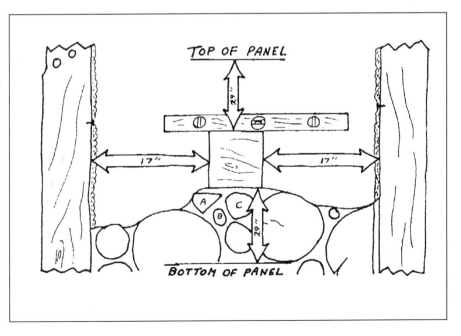

7-4. Designs incorporated into panels of cordwood.

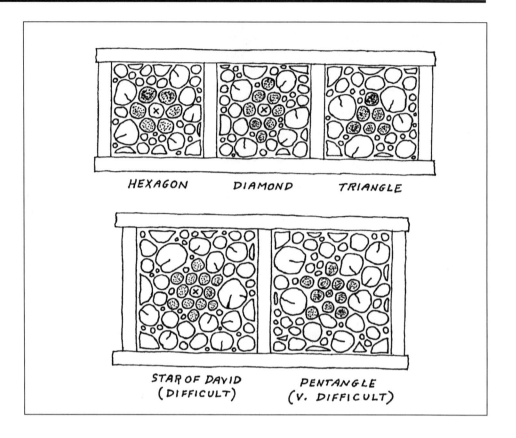

COLOR PRESERVATION

I am not an advocate of using any kind of wood treatment on the log-ends: no sealers, preservatives, varnishes, or stains. Let the wood weather on the outside. Let it breathe. This is the secret to cordwood masonry's longevity. Weathered, it will look natural and good. Some woods weather grey, others red (such as northern white cedar). We do not build cordwood masonry houses to enter into a long-term maintenance situation. Any sealer, coating, varnish or urethane will have to be renewed every two or three years. Why bother? Weathering will defeat you eventually.

On the inside, the log-ends will retain their appearance indefinitely. If they were bright and clean and fresh on the day they were put in the wall, they will be bright and clean and fresh 15 years later. If they were rough and dirty, they will not improve, unless you sand the ends with a high-speed disk sander. I use a Makita 4500 r.p.m. 5" disk sander which only cost about $65. I find it to be one of my most versitile power tools. Be sure to wear a breathing mask and eye protection. If the need for sanding is anticipated at the time of building, be careful to keep the inner surface of the log-ends all in approximately the same plane and keep all the mortar joints recessed by at least ¼". Care in these details will make for easier sanding later.

Cordwood masonry is light-absorbing. Any treatment that is applied can only darken the wall further. This is another reason that I tend to avoid treatments. Sanding, incidentally, *brightens* the wall. Having said all that, one treatment that has been used successfully is linseed oil on the interior sufaces of the wood. I have seen this first-hand at Ron and Bea Farrell's house in South Dayton, New York. The effect is to bring out the color of the wood, which is pleasing, but it does darken the wall somewhat. For this reason, the Farrells only did one panel with the linseed oil. A couple in Colorado treated their split red pine by dipping the interior end ½" into a pan of boiled linseed oil prior to laying it up in the wall. In color pictures I have seen, the log-ends look like petrified wood, quite beautiful.

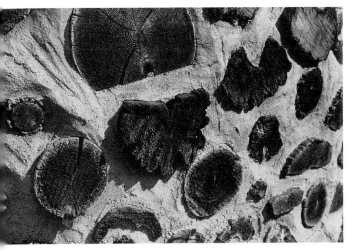

7-5. Weathered.

7-6. Bettered.

If my commentary on avoiding the use of preservatives has not convinced you, and you feel that you absolutely must put something on the wall to "protect" it, then at least be careful that you do not choose some poison containing penta, arsenic, dioxins, creosote, and the like. Apart from very important environmental questions, consider this: Even if applied to the outside only, fumes of these nasty substances can make their way through the end-grain of log-ends and into the living space. Remember, log-ends breathe much more than wood placed on side-grain.

BOTTLE-ENDS

"Bottle-ends" are simply log-ends made of bottles. They can be placed randomly in a wall, or placed deliberately in a pattern. In either case, the effect can be brilliant when sunlight or—at night—artificial light hits the panel. We have a lot of fun planning and installing these designs, and visitors seem to enjoy them, finding, I think, the same kind of appeal inherent to stained glass.

We have changed our bottle-end technology over the years as our cordwood walls became thicker. No longer do we cut and glue bottles, as previously reported. This is unnecessary and time-consuming,

and, if all the moisture is not sealed out, the bottle-end can break by moist air expansion.

The correct method of bottle-end construction—tested for over ten years without a single broken bottle—is shown in 7-7 and 7-8 The components are two bottles, a piece of aluminum printing plate or light-weight flashing, and two sturdy elastic bands. Printing plates used only once in the offset printing process can usually be obtained for free or at very low cost at your local newspaper or print shop. Wrap the printing plate into a cylinder just a little smaller than the diameter of the bottles and hold it loosely with the rubber bands. Now, simply plug the two bottles into the ends of this cylinder and, voilà, you have a very strong "spring-loaded" bottle-end. The rubber bands keep the bottles in place until the unit is laid up in the wall. Note that three inches or so of each bottle is left exposed to form a chemical bond with the mortar. (And it really does—phenomenally strong.) Here are some additional tips, garnered from our installing hundreds of bottle-ends over the years.

(1) Don't omit the aluminum, as one builder did. This cylinder is there to keep the insulation from closing in around the necks of the bottles, which reduces the transfer of light by 90 percent.

(2) Collect bottles of a height equal to half the width of the wall. Two 6" bottles make a 12" bottle-end, two eights make a sixteen, etc. One day, I was down at the local country store to buy some beer. I

7-7. Two bottles, two rubber bands, and a piece of aluminum printing plate...

7-8. ...are joined together to make a "bottle-end."

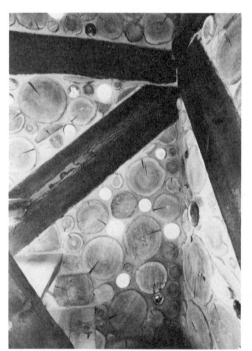

7-9. Bottle-ends by day...

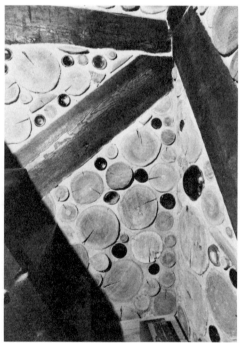

7-10. ...and by night

was measuring the heights of the beer bottles for use in our wall. The owner, after watching me for a minute or two, could contain himself no longer: "For Pete's sake, Rob, they're all twelve ounces!"

(3) Use a clear bottle and a colored bottle together to make a colored bottle-end. This lets a lot more light through. Two colored ends together are quite dark and not nearly as effective. For the most vibrant color, lay the bottle-end up with the clear end to the outside and the colored end inside.

(4) Put the bottle designs at or near eye level. They are not effective down low or very high in the wall. Plan them where the sun will hit the wall, but not too close to windows, if this can be avoided, as bright window light tends to detract from the very effective contrast between bottles and the dark log-end background.

(5) As with all special features, make a note on

7-11. Darin's bedroom panel features bottle-ends, shelves, and a fishbowl.

Rose. Avoid plastic, subject to untraviolet deterioration.

The 6"-thick interior wall panel in Darin's bedroom features a glass fishbowl in the center. (*See* 7-11.) A colored glass keepsake could be installed inside as a feature. We will be mounting a decorative glass hanging on Darin's side of the wall.

TERRARIUMS

At Log End Cottage, we laid up a one-gallon storage jar in one of the walls for use as a terrarium. Loosening or tightening the stained-glass cap regulates the air flow. Aggie Evans laid up a five-gallon glass jar in her wall for the same purpose.

STAIRWAY

This stairway looks as if it were made from log-ends. As a matter of fact, the ash steps do go right through the wall and form stairs on both sides, but the whole

the slab or use some other means of reminding yourself to install the design. Have all the bottle-ends made up and ready to go. It is so easy to forget to do something like this, and it is hard to stop work to run around collecting materials. Gather everything together the night before... even the winter before!

(6) With random designs, try to get a pleasing balance of location and color; a certain amount of "planned randomness" may be necessary. With deliberate designs, such as those suggested in 7-4, careful planning is required. But by the time you get up this high, your skills will be up to the task. One ambitious student is planning an elaborate "phoenix" design in a cordwood panel. I have advised him to lay a sheet of plywood out on the ground near the work site and "mock up" the design, standing on end the very logs and bottles which he'll lay up in the wall, and leaving the usual one-inch space between each piece for the mortar. The panel can now be laid up quickly and accurately just by taking the pieces off the bottom of the mock-up and laying them up in the corresponding wall section. Proceed in courses.

(7) The browns and greens of beer and wine bottles go very well together in a wall. We usually mix in a few clear bottles, too. Blue bottles are hard to find. Old milk of magnesia bottles are one source, and some mineral waters come in blue bottles. We recently bought 24 blue glass tumblers very cheaply at a garage sale. Glasses and jars work fine, as do odd-shaped bottles, such as Mateus

7-12. Filling in the stairway panel at Log End Cottage.

7-13. The finished stairs.

structure was built before any log-ends were laid in this panel. The steps are very heavy and the load is completely carried by the three-by-ten runners.

Figure 7-12 shows the work in progress. The third stair step is the only one of the six that is made of two separate (18") pieces, unavoidable here because of the interference of the diagonal support members. The other five are a full 45" long—18" on each side, with 9" hidden in the wall. Figure 7-13 shows the finished stairs from the inside. The outer stairs serve as a fire escape from a "tea deck" that adjoins the loft.

SHELVES

Cordwood masonry walls lend themselves beautifully to the addition of shelves. At Log End Cottage we employed two different types of shelf. Figure 7-14 shows individual slat-ends protruding 7" into the room. These make fine display shelves and are grand for candles. Another kind of shelf is also shown to the left of 7-14 where a board crosses two protruding

7-14. Display shelves.

7-15. A bottle-end sunrise brightens the bedroom on sunny mornings.

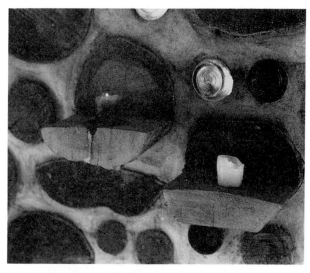

7-16. Shelves cut and split from longer log-ends.

slat ends. *Use a carpenter's level to get this exactly right*. The shelf is placed to give continuity to the adjacent panels.

At Earthwood, we used a particularly large and beautiful piece of red pine as a shelf in our bedroom. (See 7-15.) We topped the shelf with a sunrise design made of brown bottle-ends. The design was orientated to the rising sun on the longest day of the year—the summer solstice—where it lights up at dawn on or about June 21st.

Another style of shelf is shown in 7-16. This kind is made by cutting partway through an over-long log-end and then splitting off a chunk with a chisel or axe. The rough top surface can be made smooth with a circular sander. Choose straight-grained logs for this type of shelf, to avoid a twist in the split.

Related to shelves are coat and towel pegs, which are simply smaller-sized protrusions from the wall. They can be made of long narrow log-ends, or even railroad spikes, which we have used many times.

MARBLES

Kids like to put marbles in the mortar joints. So does Jaki. One of the better designs is the four-leaf clover shown in 7-17. If you look closely, you will see hairline cracks in the mortar joints around the

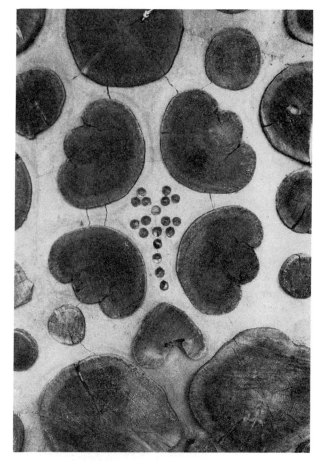

7-17. A 4-leaf clover, with marbles.

7-18. Small windows, framed with full-size two-by-two planking, can easily be built into a cordwood wall.

design. These cracks came from over-pointing. In trying to do too smooth a job around the clover design, we pulled a lot of moisture up to the surface of the mud. It shrunk, of course, when it dried. So, yes, it is possible to over-point.

SPECIAL WINDOWS

Cordwood masonry will easily support small windows floating in a panel. Illustration 7-18 shows how two such windows, framed with 2" thick planking, could be built into a cordwood wall. How about a porthole?

FANCY POINTING

We give our students free rein (almost) at our cordwood workshops. Ed Burke of Thornton, Ontario, thought that our sauna should be protected from evil spirits by a "diva" and a Hindi good luck symbol

shown in 7-19. Ed is also resposible for the portrait of Rob and Jaki in 7-20 and the apple of 7-21. Merlin Larson created the Indian face shown in 7-22.

OF MUSHROOMS, CASTLES, AND GOBLINS

All sorts of fanciful things show up in cordwood walls. And in woodpiles. Keep your eyes peeled, as well as your logs! Remember, you don't have to be crazy to build cordwood walls, but it is no handicap.

(1) Mushrooms seem to pop up everywhere in cordwood walls, and they are easy to make. This example (7-23) is, appropriately, at our "Mushwood" cottage.

(2) Goblin No. 1 (7-24). Cordwood goblins are happy goblins. This one just showed up in the wall during a workshop. Planned design? Mysterious spirit? Coincidence? Hoax? You be the judge.

(3) Goblin No. 2 (7-25). Goblin or self-portrait? Which one are we talking about? Take your pick.

(4) Special plates (7-26) are easy to set in the wall. Simply cut an inch or two off a log-end of the same diameter as the plate to be installed. Install the short

EXAMPLES OF FANCY POINTING

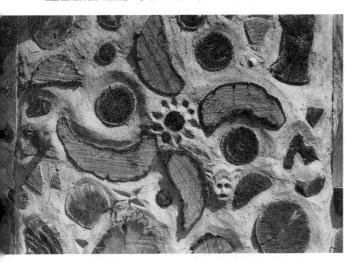

7-19. Ed Burke's "diva."

7-20. Rob and Jaki, by Ed Burke.

7-21. This apple design is in the Earthwood sauna.

7-22. Merlin Larson's extraordinary painting technique.

7-23. Mushroom at Mushwood.

7-24. This happy goblin appeared during a workshop.

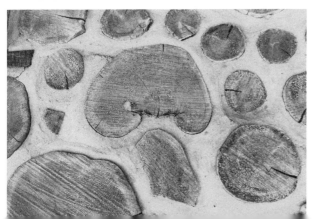

7-25. Rob Roy and friend.

7-26. This relief is based on a 14th-century ivory carving and depicts *The Siege of the Maiden's Castle*. It is set into the cordwood masonry wall at Earthwood, which shares some of the same medieval lines of architecture.

7-27. My personal favorite.

7-28. Cordwood Masonry 303.

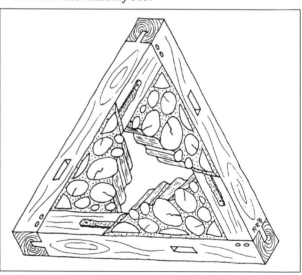

log-end recessed into the wall a bit and mortar the design plate into the space thus created.

(5) Goblin No. 3 (7-27). My personal favorite. You've got to love this fellow, who just showed up one day while we were cutting wood. Eyes, nose and mouth were natural, just parts of this old cedar fence rail which were starting to go bad. All he needed was his two front teeth, found in the woods in an ancient cow's skull.

Go ye and do likewise.

Have fun.

And when you think you've mastered all there is to learn, you can tackle the following (7-28): Cordwood Masonry 303.

TWO
EARTHWOOD

Preface to Part Two

In today's economy, there is little option regarding affordable shelter except to go out and build one's own house. Henry David Thoreau would argue that this has always been the case. Housing obtained with "hired money" can hardly be deemed affordable, if any value at all is placed on the time one spends paying off a mortgage. My wife, Jaki, and I, having put together a grubstake through the sale of a cottage which I'd renovated in Scotland, came to the United States in 1975 wishing to be free of shelter costs forever. This is a common goal of young people today, a realistic and attainable goal. However, like so many would-be owner-builders whom we meet today at our Earthwood Building School, our dream house did not encompass important economic and living considerations beyond shelter. Log End Cottage, as we called our first building effort, was little more than a strong, dry living space and a pretty fantasy.

Now, *shelter* is a "necessary of life," as Thoreau points out in *Walden*, and it is useful to know his complete definition of this:

> By the words, *necessary of life*, I mean whatever, of all that man obtains by his own exertions, has been from the first, or from long usage has become, so important to human life that few, if any, whether from savageness, or poverty, or philosophy, ever attempt to do without it.[1]

The other "necessaries of life" which Thoreau lists are essentially the things required to keep the body up to temperature and, therefore, alive: *food, fuel,* and *clothing*. Once we free ourselves from the need of spending a great deal of our time in attaining these necessities, our "vacation from humbler toil (has) commenced."[2] We are free to go about the business of living, whatever we may perceive this to be.

Unfortunately for us in 1975, and for many others today, we did not know that in the act of building our shelter, we could have been simultaneously attending to *food, fuel,* and *clothing*. And further, the building could have been much better equipped to pursue our various *home industry* and *recreational* activities (without overtaxing the carrying capacity of the kitchen table). Although we had a garden and a limitless firewood supply (a 40-acre woodlot) at Log End Cottage, we did not appreciate that the house itself can contribute to food production and preservation, as well as energy conservation. Our neglect of home industry and recreation considerations prevented the house from being a complete living place.

After two years—and the addition of six pounds of little boy—the inadequacies of the Cottage became clear. We needed a better-designed floor plan and much greater energy efficiency. We built an underground house, Log End Cave, about 120' from Log End Cottage. We oriented the house properly to the south. (Even this simple energy consideration, known to the ancient Greeks, had been omitted at the Cottage.) We used the earth to moderate our surroundings, both summer and winter. The results were impressive. The Cottage, with less than 750 sq ft of living area, required seven cords of firewood to heat per year, while the Cave, with close to 1000 sq ft, could be maintained at a much steadier comfort level on only three cords, a dramatic improvement. And a Sencenbaugh 500-watt wind-electric system provided plenty of power, so that our fuel expenses were limited to approximately $90 worth of propane gas per year. We were learning, especially with regard to alternative building methods, but still had not twinked to the area of integrative design. Although I had included a small office in order to pursue my writing career, most *home industry* wound up back on the kitchen table. And *recreation* consisted of a dartboard as the primary supplement to reading, a 12-volt television, and outdoor activities.

The greenhouse at Log End Homestead was attached to a shed 100' away, and was not designed to resist the freeze of the harsh winters of northern New York—or even the springtimes. The garden was another 80' beyond the greenhouse. Maintenance of systems requires, first and foremost, ready proximity. Out of sight, out of mind. Weeds do not stop growing while we're not looking, nor do greenhouse plants stop requiring water.

By this time, my work had evolved into developing low-cost building systems and writing and teaching about them. My books *Underground Houses: How to Build a Low-Cost Home* (Sterling, 1979) and *Cordwood Masonry Houses: A Practical Guide for the Owner-Builder* (Sterling, 1980) had become well-read resources in their respective fields. And yet I still had the nagging sense that something was missing, that there was more to building than cordwood and earth-sheltering, more even than beauty and energy efficiency.

Against Jaki's best advice—for she knows what building does to my normally cheerful personality—I decided to build again; not just a house, but a living place. Homestead activities would be concentrated near the central core. All of the various living systems would be attended to. At the same time, we would test several new concepts in the fields of cordwood and earth-sheltering. The new place was completed in 1983 (or nearly so; is ever an owner-built house truly *completed*?), and the report follows.

Although the narrative uses appropriate examples at Earthwood to illustrate certain systems, the intent of Part Two is to impart a sense of integrative design to any plan, round or square, framed or masonry, underground or on pillars.

The only "wisdom" I can claim to have learned in my 45 years on this planet is that everyone is different. There are no exclusive "right" answers with regard to housing. Build it yourself—whatever it is that you envision—and you will "meet with a success unexpected in common hours."[3] But while you're at it, why not attend to the other entrapping necessities of life, like food and fuel, and allow for other uses of the building aside from *shelter*, such as *recreation* and *home industry*? There may never be a more economical time than at the design stage of the building process.

Part Two proceeds chronologically as far as the construction is concerned, following a discussion of reasearch and planning. The chronological presentation will, nevertheless, call attention to details which must be dealt with at the early stages to facilitate easier construction or the later integration of some particular system. This is the value of the time spent planning. Many of the systems described, such as the earth tubes, the masonry stove, and the hot tub plumbing, would be extremely difficult—if not impossible—to retrofit after the main structure is complete.

About the title of Part Two: Earthwood is the name of both our home and our building school, which specializes in teaching cordwood masonry and earth-sheltered construction. Earthwood is also a clearing house for books and plans on these subjects. But, most important, Earthwood is an ongoing experiment in low-cost and harmonious living systems. There's a quiet excitement going on here, and we would like to share it with you. For information, please send a stamped envelope to Earthwood, RRI, Box 105, West Chazy, NY 12992.

Robert L. Roy
Earthwood
1984 and 1992

Top: Cliff Shockey's second house, built of recycled utility poles, Vanscoy, Saskatchewan (Cliff Shockey photo).

Middle right and bottom left: Parts of an addition built onto Cliff Shockey's second house, Vanscoy, Saskatchewan (Cliff Shockey photos).

Middle left: A circular stairway leads to the second storey at Cliff Shockey's second house, Vanscoy, Saskatchewan (Cliff Shockey photo).

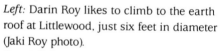

Left: Darin Roy likes to climb to the earth roof at Littlewood, just six feet in diameter (Jaki Roy photo).

Top right: Earthwood, the sauna, and the raised bed gardens, West Chazy, New York.

Middle right: A replica of a megalithic stone circle stands in front of the Earthwood house, West Chazy, New York.

Bottom right: Two of the outbuildings at Earthwood. Rohan's building, in the foreground, is 16 feet in diameter, and made of round cedar log-ends set in a Portland mortar. The office is 20 feet in diameter and made mostly of split cedar fence rails set in a masonry cement mortar.

Right: The author's office and part of the garden at Earthwood Building School, West Chazy, New York (Jaki Roy photo).
Top left: Earthwood as seen from about 25 feet up the windplant tower.
Middle left: An interior cordwood panel in Darin's room at Earthwood.
Bottom left: The Earthwood house and solar room, West Chazy, New York.

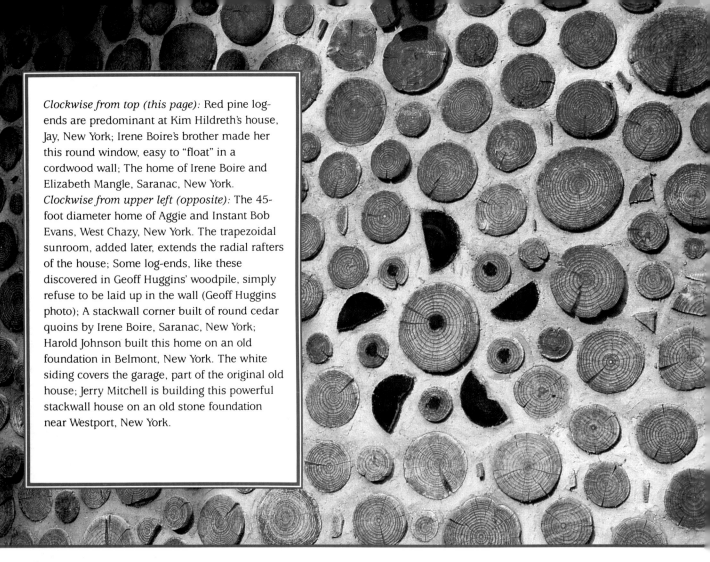

Clockwise from top (this page): Red pine log-ends are predominant at Kim Hildreth's house, Jay, New York; Irene Boire's brother made her this round window, easy to "float" in a cordwood wall; The home of Irene Boire and Elizabeth Mangle, Saranac, New York.

Clockwise from upper left (opposite): The 45-foot diameter home of Aggie and Instant Bob Evans, West Chazy, New York. The trapezoidal sunroom, added later, extends the radial rafters of the house; Some log-ends, like these discovered in Geoff Huggins' woodpile, simply refuse to be laid up in the wall (Geoff Huggins photo); A stackwall corner built of round cedar quoins by Irene Boire, Saranac, New York; Harold Johnson built this home on an old foundation in Belmont, New York. The white siding covers the garage, part of the original old house; Jerry Mitchell is building this powerful stackwall house on an old stone foundation near Westport, New York.

Above: The home of Thomas and Anne Wright, Rockmart, Georgia (Anne Wright photo).

Below: The eastern approach to "Earthwood, Jr.," built in Ashland, New Hampshire, by Paul Mikalauskas (Paul Mikalauskas photo).

Above: Kim Kildreth's beautiful home in Jay, New York, has been featured in *Adirondack Life* magazine.

Below: Betsy Pugh built this stackwall house in the town of Ausable, New York.

Top: Interior of Cliff and Jackie Shockey's house, Vanscoy, Saskatchewan (Cliff Shockey photo).
Middle: This demonstration panel was specifically constructed by Rob and Jaki Roy to provide the step-by-step sequence of photos in Chapter 3.
Bottom: A stackwall corner goes up at Jerry Mitchell's new home near Westport, New York.

8

Research and Planning

INTEGRATIVE DESIGN

We were several years in arriving at the importance of integrative design in housing. While still desiring the nearly self-reliant life-style, we approached the Earthwood project with the advantage of hindsight: We would not repeat the mistakes of the Log End Homestead. Structurally, and as experiments in low-cost and energy-efficient building, our work at Log End had been a success. But at Earthwood, we would have the opportunity to address the large picture, to integrate home and homestead in a much more efficient fashion.

No great and sudden inspiration takes place at the early stages of design conceptualization; at least, I have not found this to be the case. Influences are varied and often accidental. Earthwood as it is today is the result of several circumstances.

A ROUND HOUSE

Influenced by the work of New Brunswicker Jack Henstridge, author of *Building the Cordwood Home* (see Bibliography), I first realized the tremendous practical advantages of a round home, enumerated below. Visiting the beautiful round cordwood home of Sam Felts in Georgia removed any fear that there was anything strange or foreign about a round house. A natural comfort seemed to prevail in the buildings.

A round house of any given perimeter will enclose 27.3 percent more area than the most efficient of the rectilinear shapes, the square. Compared with the more commonly shaped house that Western man builds, where the length of the house is roughly twice the width, the gain is over 40 percent more space. Think of it: 40 percent more space with the same amount of work and materials. If a radiant heat source is placed at the center, an even heat is delivered to all parts of the house. There are no cold corners. Finally and we could not know this prior to living at Earthwood—it turns out that Jack Henstridge was absolutely right about the positive sense of comfort in a round house. Only man among the building species has deliberately sacrificed comfort for an idea of style.

About the same time that I was working on *Cordwood Masonry Houses*—1979 and 1980—I co-founded a company called Wood, Wind, and Earth, Inc., which dealt in wood stoves, wind-energy systems, and energy-efficient housing, accenting—we hoped—earth-sheltered, cordwood, and solar-assisted buildings. We signed a contract with a young couple to build the round house and began to gather the necessary cordwood to go with the rafters, already cut to accommodate the 38' 8" outside diameter of the building. Unfortunately—or perhaps fortunately, in retrospect—the contract fell through when the people decided to move out of state. Not long after this, my partner and I decided to dissolve our partnership. Part of the dissolution agreement was that I would retain the building materials earmarked for the round house.

After another "near miss" sale of the building, I decided to build it myself, still unsure if it would be "on spec" or as a new home for the Roy family. George Barber, a good friend whose genius in integrative design is displayed from time to time in this book, suggested that the building be earth-sheltered, at least on the north side. I filed the idea away without great care, as the advantage did not seem very pronounced while Earthwood remained a single-storey design.

USING MARGINAL LAND

About this time, another circumstance influenced the evolution of the project. A neighbor was interested in parting with six acres of his land, in full satisfaction of the mortgage which we had entered into together a few years before. The land consisted largely of an old gravel pit which had supplied most of the material used to build the dirt road up the hill where we live. As the gravel layer was not deep, excavation extended laterally until it consumed most of two acres.

Our ideas are built upon influences from the variety of inputs which bombard our experience. For me, these influences most often consist of other people— or books, which are simply the reports of other people. Rare is the completely original thought. One line of thinking—publicized by but not original to fellow underground advocate Malcolm Wells—is that it is high time for man to reverse the destructive process of taking all of nature's best land for the purpose of building houses. After a successful experience of building on land raped by man, Wells now says,

> When we tell our clients how to find choice building sites, we always urge them to pick the *worst* ones, not the best, as we were taught to do. Now people can see for themselves how easy and how gratifying it is to restore a bit of this trampled planet.[4]

We decided to take this old gravel pit, the visual and ecological weak link on the beautiful hill where we live, and return it to living, growing, and useful production. Even construction of a tennis court, difficult to justify in meadow or woodland, would help settle the sandstorms at the pit, while supplying a venue for healthful recreation. As of 1992, we have reclaimed much of the pit with raised bed gardens, lawn, and a megalithic stone circle. We still haven't built the tennis court, but haven't given up the idea entirely.

LEGAL DOCUMENTS

My neighbor and I saved time and money by typing our own deed description on a standard deed form (checked for accuracy by surveyor George Barber) and completing our own Satisfaction of Mortgage. There was no need for either party to retain a lawyer. All a conscientious lawyer will do is to consult with a surveyor or title company to make sure the deed is properly described and has no encumbrances. Why not save $100 to $200 and eliminate the middleman? Legal blanks are available at stationery stores for less than a dollar. Do not be put off by the admonition that: "This is a legal instrument and should be executed under supervision of an attorney." This is like saying that a home should only be built by a contractor. If there are truly legal problems connected with the deed transfer, and not just title-description problems, a lawyer should be consulted of course.

THE EVOLUTION OF DESIGN

A house design is best allowed to evolve at its own rate. I do not intend to be mystical when I say that much of this evolutionary process takes place in the subconscious, in dreams, perhaps, or during those few lucid moments between wakefulness and sleep. And many problems in design seem to solve themselves, although I expect that what really happens is that passing time enables the designer to gain a new perspective. Flash-in-the-pan special features which occupied the imagination so intently at their inception are seen in a more practical light three months down the road. Changes in thinking occur almost imperceptibly. I cannot even recall when Jaki and I first made the decision that Earthwood would be our home. Perhaps there was no decision, just a realization. But I do know this: When we left for Scotland just before Christmas, 1980, for a five-week visit, we knew not only that the house was for us, but that it would be a two-storey round house with two other nearby round outbuildings.

TWO STOREYS

A constant part of design is—or should be—costing analyses. No sense designing the unaffordable. The materials we had were right for a 38' 8" diameter round house. A rough estimate of the materials' cost of the one-storey design was approximately $12,000. A good method of stimulating the imagination in house design—or in writing science fiction, as John Wyndham was fond of saying—is to ask, "What if . . .?" I asked myself, "*What if* the house had *two* storeys? How much would this cost?"

I totalled it up. To add a second storey of the same size would require about five extra cords of wood for the log ends, a few extra posts, floor joists and 1000 sq ft of flooring, a stairway, and some extra internal wall framing and covering, including a few more doors and windows. The foundation and earth-roof structure would remain constant, as well as all the ancillary systems, such as the septic tank, driveway, and wind plant. I figured the cost of the second storey to be about $4000, just $4 per sq ft, compared with $12 for the first floor. It seemed too cheap to turn down, and provided scope for such integrative design features as a hot tub, a pool table, and a reasonably large office. Here would be a house we could grow into.

The two-storey line of thought moved in another direction besides economy. Warm air rises. This simple law of physics was known by early man tending his fires thousands of years ago and by termites millions of years prior to that, making use of the principle in their convection cooling systems, as I shall show later on. The concept was rediscovered—for us—at Log End Cottage, where in winter the sleeping lofts averaged 10° F to 15° F higher than the lower floor.

Let's use this warm air twice before it escapes us altogether. Let's put the main heat source at the center of the lower storey. This idea worked particularly well with the round design. All points on the inner surface of the exterior wall would be equidistant from the heat source. There would be no cold corners, apart from those we might wish to create deliberately through the use of interior wall insulation. Many people say—not quite accurately—that heat rises. This is not strictly so. Warm fluids, such as air and water, do rise, because they expand and get less dense and, therefore, float to the top. Radiant heat travels equally in all directions from the heat source, including down. If a radiant-heat source, such as a wood stove, is located at the center of a round building, all parts of the wall on an uninterrupted line from the stove will be treated to equal radiation. Then, if these external walls are built in such a manner that the heat can be stored, there exists a great potential for maintaining a steady temperature at the desired level.

THE STRUCTURAL PLAN

Part of our trip to Scotland was spent at the home of Jaki's parents, where I had plenty of time to work on the Earthwood design. Jaki's father had several excellent drawing tools, such as a scale rule, compass, and square.

At Log End Cave, I had learned the importance of integrating the floor plan to the structure, rather than working the other way around. This strategy keeps the plan simple, and, therefore, easy and economical to build. For example, the most economical means of supporting an earth roof—and the most esthetic—is by utilizing a properly engineered post-and-beam and plank-and-beam framework, which is left exposed on the interior. The power and strength of a building thus constructed can be felt and enjoyed by the occupants and visitors. When it comes time to subdivide space in this sort of building, rooms are best created with their walls rising up to meet the underside of the heavy timbers used for girders and rafters. This strategy lends strength to the roof structure, while avoiding the problems connected with just missing the main timbers: fitting a wall around exposed rafters, and creating awkward spaces conducive to cobwebbing. Esthetically, just missing a support timber with an internal wall is most disturbing.

The structural plan of the house was determined by the need to support a roof load of approximately 150 lbs per sq ft, about three times stronger than a normal roof structure in our area, which is designed for a snow load of 50 lbs per sq ft. The extra allowance, of course, is the weight of approximately 10" of earth and stone on the roof. (I will not interrupt the narrative at this point to wax eloquently on the benefits of an earth roof, but, instead, invite the reader to consult Chapter 19, the paragraph entitled "Earth Roofs: The Trade-offs". Suffice it to say that once a builder becomes involved with earth roofs, it is very difficult indeed to return to a dead roof design.)

As stated, the basic structural plan of the house followed a design created for *Cordwood Masonry Houses*, but it was necessary to make some adaptations in order to accommodate some desired floorplan features. A simplified version of the final Earthwood framing plan is shown in 8-1. The primary change in the plan is the elimination of the northernmost post in the octagonal post-and-beam framework. This accomplishes two purposes. Firstly, the elimination of the post makes space in a downstairs recreation room for either a billiard or Ping-Pong table. Secondly, the two upstairs bedrooms naturally fall in on the north (cool) part of the house, each with a square corner to facilitate the placement of an easily accessible bed. These are the only square corners anywhere at Earthwood, and, truly, the only ones necessary.

THE KRAAL

In the late '60's, 1 took a six-month tour of Africa, and saw thousands of round houses, built mostly by natives, but also by white settlers. I have been attracted ever since to the social and architectural possibilities of round houses, especially when clustered

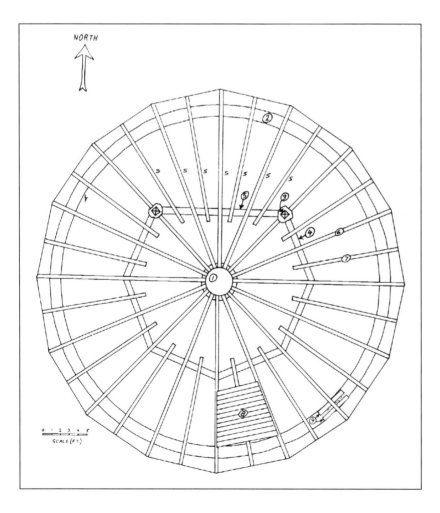

8-1. Earthwood framing plan.

1. Stone heat sink: 4' diameter upstairs, 5' diameter downstairs
2. 16" cordwood masonry wall
3. Post locations, seven in all
4. Girders, eight-by-eight best Douglas fir or equivalent
5. Special girder, ten-by-twelve clear oak or equivalent (spans 15')
6. Primary rafters (five-by-ten red pine or equivalent) or floor joists (four-by-eight)
7. Secondary rafters (five-by-ten red pine or equivalent) or floor joists (four-by-eight)
8. Two-by-six tongue-and-groove planking
9. Two-by-six plates

S = Special rafters for greater spans, six-by-ten red pine or equivalent

   Floor joists at these locations can be four-by-eight.

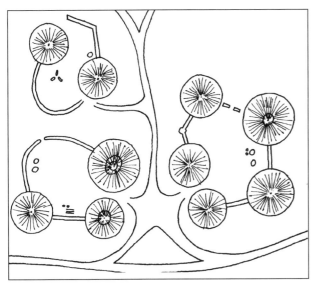

8-2. An aerial view of an African village of kraals.

8-3. An early drawing of Earthwood by Jaki Roy.

together in various ways, such as in a Zulu village. (*See* 8-2.)

Besides structural plans (both framing and cross-sectional renderings) and a floor plan integrated with the structures, I took the time while in Scotland to make paper scale models of the main building at Earthwood, as well as the two satellite structures. I used a scale of ⅜" to the foot in my flat drawings, and a ¼" scale for the models. With little roof caps on the models, it was easy to visualize the juxtaposition of the three buildings in various ways.

The plan evolved into something very close to Jaki's line drawing, reproduced in 8-3. If Earthwood were an African tribal chief's kraal (cluster of huts), the main building might house the number-one wife. Number-two wife would be in the secondary building, while number-three wife and the goat got the smallest hut. I was not able to get planning permission for this arrangement past Jaki, but we had perfectly good alternative uses in mind for the other buildings, anyway. The second building would be a storage shed, and would not be heated. Every homestead has need of lots of storage area, and, in most cases, tools and other homesteading items can be stored without protection from freezing. Eventually, the second building became my office-workshop and we built yet another 16' diameter building as a storage shed.

The smallest building is a wood-heated sauna.

There are two primary advantages to having a sauna in a separate building. One, the house is not overheated, an especially important consideration during the summer. Two, there is a pleasant feeling of "getting away" to a different place, a world apart, allowing the bather to escape any and all reminders of everyday cares—even though the sauna is only 15' from the house.

OUTBUILDING DESIGN

The outbuildings were designed with two primary considerations in mind: their function, and architectural harmony. Construction of cordwood masonry walls with earth roofs guaranteed a harmony of materials, but we also wanted the relationship between the sizes of the three buildings to be pleasing. With a bit of pencil-pushing and erasing, we arrived at a kind of geometric progression that made sense with regard to the function of the buildings and the structural requirements of both cordwood masonry in a round design and the need for a strong support system for the earth roof.

A 9' internal diameter for the sauna seemed to be the optimum size for placement of the stove and benches, without the building being too big to heat. Again, the heating advantages of the round shape come into play. The cordwood wall would be 8" thick, half the wall thickness of the house. Our experience with a rectilinear earth-roofed sauna at Log End showed that this was ample wall insulation, and the walls would not be so thick that there would be a great disparity between the inner and outer mortar joints, even at this relatively small diameter. Finally, a heavy earth roof could be supported without the use of an internal post.

To lend a sense of balance to the complex, the shed would be geometrically between the sauna and the house in size, an 18' internal diameter (20' o.d. with the 12" cordwood walls). At this size, one post is necessary to help support the roof load, located right at the center. The 254 sq ft of storage area would be ample for our purposes, although we long ago learned that no matter what the size of storage space available, it will get filled up. Thus, the amount of storage space influences the quality of one's junk collecting, as well as the quantity.

FLOOR PLAN—UPPER LEVEL

Jaki and I spent some time on the floor plan (8-4). The first decision was to design the upstairs space so that it could exist as a simple dwelling apartment, independent of the downstairs. This was not difficult, as the original plan had a single-storey building in mind. What we ended up with was actually quite similar to the Log End Cave floor plan, which we had liked, except that the almost square shape had been turned into a circle—as if graphic artist M. C. Escher had performed one of his space-twisting tricks with the Cave design. Because of the earth berm on the northern hemisphere, it is possible to enter either storey directly, without having to climb stairs. But the door downstairs, as well as openable windows in all rooms, eliminates the need for a second entrance upstairs.

Entry upstairs is by way of a combination mudroom-wood store, the firewood to be used in the cooking range. A mudroom serves as a thermal break between the true interior of the house and the great outdoors, and affords a place for boot removal and the parking of hats and coats.

Bedrooms, which should be somewhat cooler than the public rooms of a house, are placed in the northern hemisphere. Earth berming tapers upward to within 2' of the roof rafters, but leaves room for medium-sized crank-out windows further from the northernmost wall point. Our original plan called for a small skylight in each bedroom for additional light, but we did not install them during roof construction because of the lateness of the season. We now feel that they are unnecessary.

The nook serves as Jaki's private offfice and sewing area, and affords access to both the master bedroom and the bathroom without the doors of these rooms opening directly onto the main living space. This is a great advantage in making use of space in a house. A similar idea could have been employed on the other side of the house, had we eliminated the wood store in favor of a second nook. Both the mudroom and son Rohan's bedroom, then, would have opened into the nook instead of into the main living area. But our kitchen

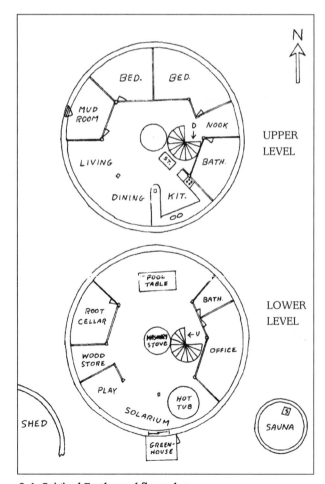

8-4. Original Earthwood floor plan.

firewood would then have had to be stored outside, for lack of room. We opted for dry warm firewood, and do not regret the decision.

The bathroom is ample in space, even for the inclusion of both the hot and cold water tanks, components of the special pedal-powered water system described later. The bathroom is convenient to the master bedroom, and just down the hall/library from Rohan's room. There is a half-bath, with shower, downstairs.

The main area at Earthwood is large and open-planned, and consists of the kitchen, the dining area, and the sitting room. A sense of separate rooms is provided by the two posts, and the heavy girders and rafters above, as well as the multisectioned plank floor below. The barlike countertop built around one of the posts serves as a divider between the kitchen and dining areas, while providing ample storage for each. An old oak church pew, 8' long, fits perfectly between the other post and the external wall, partitioning the dining and living areas. A curtain of beads hanging over the back of the pew would give further definition to the conversation area, but, again, this is an option which we have not utilized. The airtight cookstove (an Oval by Elmira Stove Works of Ontario) is centrally located and heats the upstairs in the wintertime, gently assisted by the masonry stove described later.

The hall-library is large enough to accommodate heavy shelves which we use for our personal books and magazines, as well as those we sell by mail order and at our building school. The telephone is also centrally located in this area, just at the top of the circular stairs.

It can be seen that the upstairs portion of Earthwood could be used as a complete house for a small family requiring only two bedrooms. The effective floor area would be 1018 sq ft. A single-storey plan of this kind would benefit from additional space gained by the elimination of the stairwell. My advice for a second entrance—for safety and to meet code—would be to place it in the nook, although this upsets the use of the nook space somewhat.

FLOOR PLAN—LOWER LEVEL

Downstairs space has a completely different character and use. It is truly an earth-sheltered space, as opposed to a cellar, as care was taken to provide plenty of natural light, walls are bright and properly waterproofed, and a continuity of the exposed plank-and-beam architecture is maintained. The floor is concrete, but will be covered in the future with a high-quality floor covering or slate, when we are able to afford it.*My office is large enough to double as a guest room, and we have used it as such on several occasions. It is also large enough to store my work-related books, magazines, and plans; and it is possible to draw and write without having to clear the desk with each change. All of these functions were pushed beyond reasonable limits at my tiny office in Log End Cave. Others may use this space for purposes pertinent to their work or home industry, such as for a workshop, studio, or beauty parlor. The half-bath is adjacent to the office.

Because of the earth-bermed design of Earthwood, the northern hemisphere of the lower storey gets a minimum of natural light. Walls are painted white to maximize the light which does come in from the south-facing windows. But this lack of natural light does not affect the use of a large part of this space as a place for indoor recreation, as artificial light is most often used for this type of activity anyway. Specifically, the space panders to my longtime fascination with using a long tapered stick to punch little plastic balls into leather-lined holes. An inordinate amount of space is required to entertain this folly, and I even had to design a post out of the structure to accommodate the billiard table. A post rising out of the center of the table tends to spoil a lot of shots. The solution, a very heavy oak beam, is discussed later. Jaki's preference in table sports is Ping-Pong. A table-tennis surface can be placed directly on the heavy pool table, but the plywood surface (5' × 9') will have to be specially ordered and the surface height will be 2" or 3" high-

---

* Items of this nature should not delay entry into your own home. A completely finished home—floor coverings, new furniture, and all—is a pleasant fantasy for many, but very expensive to implement. While waiting to be able to afford the whole package, the buyers are paying for their shelter somewhere else. If a loan is involved, it will be very much higher if all finished items are to be financed. A $20,000 loan may end up costing several times more after the dust clears than a $10,000 loan, and take several times as long to pay back. Jaki and I have a policy of not buying what we can't afford. Federal and state governments ought to try this approach.

er than regulation. I have not found that this affects the enjoyment in any way.

If the large space required for pool or Ping-Pong (about 250 sq ft) is not required, the effective space at Earthwood suddenly increases by 12 percent. A four-bedroom house with two full baths is easy to accommodate; but, the floor plan will have to be juggled around if natural-light requirements and building codes concerning egress from bedrooms are to be accommodated. Bedrooms, in this case, would be placed in the southern hemisphere, and other spaces not requiring natural light would be concentrated to the north. The hot tub could be placed there, for example (although I would not consider this to be a preference), as well as rooms such as a laundry, office, workshop, storage area, bathroom, or utility room.

Our utility room is located in such a space, with no natural light and but one way of entry. The room serves as a root cellar for food storage, and houses our batteries for the wind-electric system. In addition, our pedal-powered water pump is located there, as well as too much stored junk. With food, water, and electricity available, as well as concrete block walls on all sides and an earth roof covering the entire building, it was also considered at the design stage (with no "end" of the Cold War in sight) that the utility room would provide shelter for a few weeks in case of nuclear attack on the Plattsburgh Air Force Base, 15 miles away. I don't think anyone truly knows the effects of modern-day nuclear weapons. The room may do little more than allow the occupants a look-see—like the February groundhog—to better decide if an attempt at further survival is worthwhile. An all-out nuclear confrontation is probably not worth surviving, although I am not completely sure of this. And, as a friend once said in a moment of lucidity: "The two really significant points in man's history are his beginning and his end. I wasn't here for the beginning. I sure don't want to miss the end."

A wood store, accessed by a removable chute, occupies the space next to the utility room. Firewood is convenient to both the masonry stove and the special stove used to heat the hot tub.

The hot tub itself is located on the south side of the building (if a round house can be said to have sides), in order to take advantage of light and views through the windows and the sliding glass door which leads to the greenhouse .

The remaining space in the southern hemisphere of the lower level is bright and cheery, and was originally designated as Rohan's private play area, where he could leave toy trains set up and entertain his friends away from the mainstream of activity.

Each of the downstairs rooms is vented by connection to the earth tube system, which is explained in detail in Part Two, along with the masonry stove, the bicycle-pump water system, and the cordwood hot tub.

I have taken the time to describe our floor plan not to try to convince the reader that he should adapt it to his needs. On the contrary, everyone's needs are different. Rather, my intent is to show some of the kinds of considerations and tough decisions that come into composing the floor plan. The function of a room will largely dictate its location with regard to access and light, for example. And the importance of designing a simplified structural system, to which the floor plan is fitted, cannot be overemphasized. This is particularly important when dealing with exposed floor joists and roof rafters.

TEN YEARS LATER

The previous few pages on floor plans at Earthwood were written soon after we moved into the home in 1982, and are a true account of our plans and use of the space. I leave the account as originally written, because it gives a good sense of the kinds of factor that must be weighed in choosing a floor plan.

In 1985, Darin Scot was born at home. (Two midwives were present; one of them, however, was Jaki herself.) The way we used the space soon needed an overhaul. But, over the years, we have found the house to be very accommodating to floor plan changes. The first big switcheroo was Darin moving into Rohan's bedroom upstairs, Rohan moving into my office downstairs, and my office moving out to the shed, where I am now pounding out these words. Later, a new 16-foot-diameter shed was built as a workshop project.

The second big change came in late fall of 1991. The boys were growing up, and the house seemed to get smaller. Rohan needed his own space to listen to his wild teenage music, watch the stupid programs he likes, and entertain his weirdo friends. (Just kidding, Ro.) And we wanted a guest room upstairs for

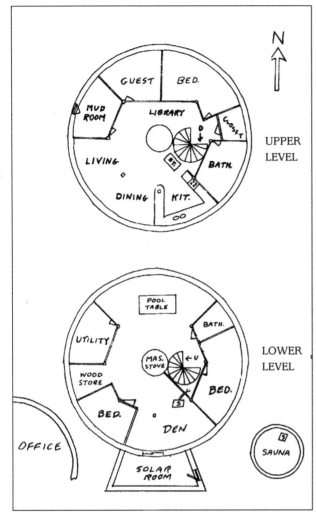

8-5. Earthwood, 1992.

Do we miss the hot tub? Actually, we don't, but this is because we took out a family membership with a local health club, where we play squash and use all of their facilities without all the work and care that comes from having one's own tub.

RESEARCH AND DEVELOPMENT

These were the things that Jaki and I hashed out while in Scotland. There remained a great deal of research upon our return home into such things as masonry stoves, earth tubes, hot tubs (especially heating one with wood), and other systems which cannot rely on "regular" 115-volt alternating current. With one exception, which follows in the next paragraph, I will not bother to relate the history of our researches, as the results are of much more importance.

There were two areas about which I did not know where to begin to get authoritative information, specifically: (1) the application of a waterproofing membrane directly to a wooden deck substrate, and (2) the effects of water runoff from a freestanding earth roof. At Log End Cave, the expanded polystyrene insulation was placed below the membrane. While this has not caused any problems, I am now convinced that the prevailing thinking in the earth-sheltering field is the correct line: the insulation should be on the cold side of the membrane. By this method, the membrane is protected from freeze–thaw cycling as well as from work boots during the placement of the earth.

We conducted our tests with the membrane and the freestanding earth roof at the cordwood sauna, described in Part One. The waterproofing method described in *Underground Houses*—built up layers of black plastic roofing cement and 6-mil black polyethylene—has done an excellent job of keeping Log End Cave dry for five years. However, the procedure, while inexpensive, was hard, dirty work and unbelievably time-consuming. It was not until the Cave was finished that I heard about the W. R. Grace Bituthene® waterproofing membrane, which provides equal or better protection with one-twentieth the work.

The earth roof at the Cave meets with the earth berm which shelters the east and west walls. A pedestrian walking over the roof has difficulty in

my mother's visits, a room which could double as a kind of media room. So the hot tub was removed, which took up a lot of space downstairs, and a second living room, the kids' living room, was installed in the old hot tub space.

We put a beautiful slate floor down throughout all of the "public rooms" downstairs—this will be described later—and Darin moved into a bright new bedroom. The upstairs "nook" became a walk-in closet, Jaki's desk landing up in the new guest-media room. Incidentally, both downstairs bedrooms are code-worthy, having opening windows and plenty of natural light. Even with the pool table still occupying a lot of space, the house easily accommodates four bedrooms.

establishing when he is actually over the house. A great advantage of this design is that rain runoff is natural and slow. Excess water on the roof percolates gently through the sand backfill against the house, finding its way finally to the French drain (footing drain) system below.

The waterproofing detail where the roof meets the wall was quite simple at the Cave, as both the roof and the wall were flat planes. Any sheet membrane easily bends around the line of intersecting planes. If the sod roof at Earthwood were to join with the earth-bermed wall, however, the waterproofing detail would be a nightmare. The roof is composed of 16 flat facets, each in different planes, much like the blunt end of a diamond. The wall is part of a cylinder. Any intersection between roof and wall, therefore, would be extremely complicated, and impossible to waterproof using sheet membranes. (Liquid polymer membranes are expensive and would also be virtually impossible to do properly at the critical roof join.) The only reasonable approach at Earthwood would be a freestanding earth roof. But what would happen during a heavy rain? Would mud flow off the building?

The answers were provided at the Log End sauna, with its 8"-thick freestanding earth roof. We observed that no runoff was visible until the earth was completely saturated, and, then, that most of the runoff occurred at the bottom of the single-pitched roof. Actually, in the years that the sauna has been standing, we have found that serious runoff is a rare occurrence. Our climate is one of moderate rainfall, and rain usually stops before the roof is saturated. In a wet climate, such as the Pacific Northwest, runoff would be much more frequent, perhaps even regular, and gutters would be necessary. We installed gutters over the deck portion of Earthwood in 1984.

We set the retaining timbers at the sauna on a ¾"-thick bed of hay to act as a filtration mat. This worked quite well, in that very few soil particles filtered through after the first couple of months. The water was colored—by the hay, it seems—but not dirty. We employed a similar method at Earthwood, but found that an important adjustment was necessary, as described at the end of Chapter 19.

PLANS AND MODELS

As stated, the really important plans—the ones we build from—are the structural plans. Primarily, these consist of a rafter plan, a footing plan, and a cross-sectional plan. Floor plans can be integrated with the rafter plan later. In many two-storey houses (including Earthwood), the floor-joist system is virtually identical to the roof-rafter system. There may be members missing to allow for a stairwell, but usually the plans can be superimposed upon each other and used for both purposes. This keeps the major struc-

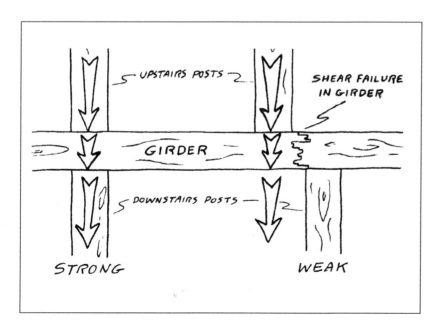

8-6. The arrows indicate lines of thrust from the roof.

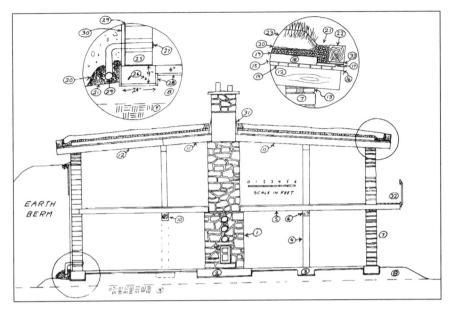

8-7. North-south cross section of Earthwood (greenhouse omitted)

11. Five-by-ten rafter, red pine or equivalent
12. Six-by-ten rafter for longer span, red pine or equivalent
13. Two-by-six wall plates
14. Two-by-six tongue-and-groove decking
15. W. R. Grace Bituthene 3000 or 3100 waterproofing membrane
16. 10" aluminum flashing, under Bituthene, folded over planking as a drip edge
17. 1" rigid foam as protection board
18. 4" Dow Styrofoam®insulation
19. 6-mil polyethylene
20. Hay or straw layers
21. No. 2 crushed stone
22. Railroad tie or eight-by-eight pressure-treated timber

1. Stone mass/masonry stove
2. Stone mass footing
3. Pillar footings, seven in all
4. Eight-by-eight post
5. Four-by-eight floor joist
o. Eight-by-eight girder
7. 16" thick cordwood masonry wall
8. Compacted sand pad (could be good gravel or crushed stone)
9. Undisturbed earth, no organic material
10. Special ten-by-twelve clear-oak girder (or equivalent) for 15' span

23. 8" of topsoil, planted to grass or wildflowers
24. 4" perforated flexible drain with nylon or fibreglass filtration sock
25. 8" concrete corner block, first course bedded in mortar for levelling
26. No. 5 (⅝") reinforcing bar
27. 4" ABS plastic pipe (earth-tube inlet)
28. 1" Dow Styrofoam®
29. W. R. Grace Bituthene membrane
30. 2" Dow Styrofoam®or equivalent
31. Galvanized flashing cone
32. Walk-around deck with railing
33. 1" pressure-treated shim.

tural components exactly over each other for much greater strength. Because of our very heavy roof, the cross-sectional dimensions of the rafters are much greater than those of the floor joists: five-by-ten and six-by-ten, instead of four-by-eight. Posts on the upper storey are always right over the lower-storey posts. The roof load, then, is carried through the upstairs posts, through the girders on compression only, and through the downstairs posts, finally supported by substantial pillar footings beneath the concrete floor. This is critical. If an upstairs post just misses being aligned with its brother below, as is illustrated in 8-6, the likelihood of shear failure in the girder is very great.

The cross-sectional plan (8-7) enables the designer to plan ceiling and rafter heights, the location of wall plates, door heights, roof details, and footing construction. With a rectilinear house, it is a good idea to draw elevationals of the four sides. These plans enable the designer to see what the house will look like, while helping to solve some construction details such as window and door placement, plates, snowblocks or soffits, and the like.

With a round house, a cross section through the center is of great use, but elevationals are not. The side of the house is not flat in the first place, rendering a transposition of detail from a curved wall to a piece of paper virtually impossible and even the best

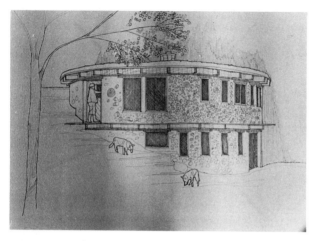

8-8. Architect Marc Camens's perspective drawing of the Earthwood house.

gives a circumference of (121.49') 121' 6". At the quarter-inch scale, the paper cylinder would be rolled from a piece of paper 30⅜" long. Standard 6'8" doors determined the height of the building, which was 15' 6", or 3⅞" on my scale model. I drew all the exposed rafters and plates first and labelled the four rafters corresponding to the primary compass points to aid in orientation.

WINDOWS

We already owned twenty uniformly sized 1"-thick thermopane window units, each measuring 15½" × 40½". We'd bought the units a couple of years earlier for $75 the lot. Apparently, they had all been cut an inch short for a bank job. We bought them at low cost as "miscuts," also called "shop units." They were a perfect shape for Earthwood: tall and narrow. Windows wider than 4' do not work in well with the round house, as they cut across the perimeter quite a bit. We had enough of the window units to experiment with different placements, upstairs and down. The odd size would not matter, as we would build the frames to fit the units. Another advantage to the narrow configuration was that they kept the walls very strong. It was possible to design Earthwood so that the roof load would never be transmitted directly through the rafter at a point right over a window. Windows and doors are always between the rafters (or floor joists). We would need other windows, of

efforts practically useless. An artist's rendering, such as architect Marc Camens's drawing (8-8) may be helpful in impressing a bank officer or a building inspector, but will not help a great deal during the construction. In a round house the best way to plan the wall details—such as the location of doors and windows—is to build a simple scale model. This can easily be done with paper, as shown in 8-9 and 8-10. If possible. keep the scale of the model the same as the scale for the cross-sectional plan. I used one-quarter inch to the foot for our Earthwood model. The outside diameter of the building was to be 38' 8" (38.667'). Multiplying the diameter by π (pi = 3.142)

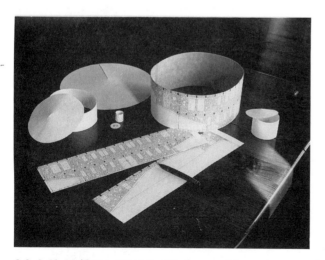

8-9. 8-10. Making a paper model of a round house is easy and very useful.

course, especially opening units, but the twenty mis-cuts helped to establish a pleasing motif to the house, not unlike a medieval castle.

Windows were roughed in on the floor plans to make sense with the room usage, and then trans-posed to the scale model. When we arrived home from Scotland, we checked into the sizes of the vari-ous opening window units available, and chose mates closest to the twenty fixed units we already had. We designed three double units having one opening and one fixed window each. We also chose opening units for the bedrooms, the sewing nook, my office, and Rohan's school area. Two other large shop units, each about 3' x 5' were earmarked for the living area, completing the windows. Because of the 16" wall thickness. the window wells in the living room make excellent seats.

SYMMETRY

Early on in our pencil pushing and line scrubbing, it became apparent that a pleasing symmetry could be maintained with just a little effort. The south rafter became the line of symmetry, and the predominant windows equidistant from the center and the rafters. The line of symmetry would also intersect a sliding-glass-door unit, which would open into a green-house on the lower storey. A second-storey deck would broaden out to form a sitting area just above the greenhouse.

Viewed from the south, the main building at Earthwood is quite symmetrical. As the house recedes to the east and west, the symmetry is lost, but this is not noticeable from the front. On the west "side," there are the two large windows and the main door. On the east, there are opening window units for the bathroom and nook.

Symmetry in a house is not an important end in itself, unless it happens to satisfy some particular architectural aspiration of the designer. But with a cylindrical house employing a radial rafter system, there is an automatic symmetry which is pleasing to flow with, and which aids in the simplification of construction. The different sizes of the outbuildings keep the kraal design of Earthwood from being fas-tidiously symmetrical.

THE DESIGN STAGE ENDS

The main work of design carried on over a period of six or eight months, a long enough time so that some problems managed to solve themselves, and bad ideas fell by the wayside through attrition. Most, if not all, structural problems were solved. Stress-load calculations were made for the roof-support system. Footing dimensions were planned for equal settling of walls, posts, and the central masonry mass. Details of window-frame construction were worked out, as well as their installation with regard to lintels. We cannot solve all problems at the design stage, but we can solve most of them.

The actual number of hours involved in planning and research could probably have been fitted into five weeks of steady work, but I do not believe that the results would have been as good. Oldtime mas-ter-builders allowed heavy timbers to season a year for each inch of thickness, believing that rapid dry-ing in kilns "crisps wood to death." I believe that the same sort of thing is true with house designs. The ideas must be allowed to season slowly.

PLANS

The Earthwood project is used as the model for Part Two, in order that real firsthand experiences can support the techniques described. In all my writings in this field I have endeavored to report accurately my own successes and mistakes, and not merely to comment secondhand on the experiences of others. That the reader may wish to design and build a house of a different shape or appearance from Earthwood does not diminish the importance of the specific systems and techniques described. The order of construction, the construction details, and the description and installation of the various inte-grated systems are the important matters here, not the specific design. If the owner-builder wishes to use the Earthwood design, or an adaptation of the actual plans, I certainly have no objections. Detailed plans, drawn and stamped by architect Marc Camens, are available from Earthwood Building School, RR 1, Box 105, West Chazy, NY 12992.

9
Site Work

THE CUT-AND-COVER METHOD

All too often in American construction, a contractor tears up a living section of the planet's surface with heavy equipment, builds a house on the moonscape thus created, and then attempts to restore the land to a living green space again, under the guise of landscaping. This is the "expedient" way to go about the job, perhaps even the "economic" way. And very often the final product can be considered a cosmetic success, even though many people's sensibilities to this sort of thing may be offended. My main objections to this strategy are twofold: (1) We cannot improve on what nature has taken hundreds of thousands of years to create, especially the ecological balance; (2) in most cases, the topsoil, one of our three most valuable resources (along with water and air), is destroyed by homogenization with the subsoil. This can be minimized. The owner-builder and the heavy-equipment operator can carefully scrape the topsoil to the edge of the building site, exposing undisturbed subsoil for building upon. They can use the valuable "black gold" on the roof of the house or to improve a garden space. The careful approach may indeed take more time—not necessarily more money—but the rewards are worth the effort.

In our case, we eliminated the first step—the destruction of the land—as this had already been accomplished by the roadbuilders of previous generations. We could build with a minimum of site preparation, confining our considerations to those structural. We could minimize the reclamation necessary, especially in the area near the house, by building into the north edge of the gravel pit, and this we attempted to do. This would give the house a southern exposure, while the woods to the rear would offer protection from the harsh north winter winds. Unfortunately, problems were encountered, partially of my own making.

THE WELL

1980 was one of the driest years on record in northern New York. Dug wells which had given consistent service for 50 years were drying up. By November, the water table was so low that the 100-year-old, stone-lined well at Log End was giving us trouble. This was the perfect time to dig a well! If water could be found in these bad times, a good well indeed would be the result.

WITCHING

I contacted old Bill Rivers of Beartown, who, by the little understood but widely appreciated method of water witching, had found water successfully for a number of people in the neighborhood. Using a pair of welding rods bent at right angles and inserted into hand-held short sections of copper tubing, Bill walked back and forth across the area where we would have liked to find water, just to the west and north of the house site. Bill was practiced in holding the copper tubes upright, the bent welding rods

142

pointing ahead of him, parallel to each other. When the rods began to turn towards each other, Bill knew that he was approaching a "vein" of water, as he called it. When they crossed, Bill was directly over the vein. After a number of trials with the same results, Bill was able to establish a vein running very nearly along the banking formed by the termination of the gravel pit. We chose the "strongest" point on the line for our well, and thanked Bill with confidence. And then I made my first mistake.

Ed Garrow, a very careful equipment operator with over 40 years experience, came in early December to dig the well, using a giant backhoe mounted on tracks, capable of digging a hole 19' deep. I showed Ed where the well was to be dug, but I had no idea of the size of the hole which would be created in the event that Ed had to "dig himself down" in order to gain greater depth. We approached the well site from the east—that is, from the direction of the house site—and failed to strike water at 19'. We decided to "dig the machine in." The backhoe required a huge hole to sit in, as well as a ramp to drive down into the hole. My mistake was that I allowed this hole and ramp to be dug on the house side of the well, little realizing the size of the crater which we would create.

At 23' of depth, Ed could get no deeper from his position 4' below grade. We stood together dejectedly on the edge of the hole and then Ed said in his slow, methodical manner, "Look, you've got water." At the bottom of the great hole I saw what looked like the puddle from a spilt cup of tea. "*That's water?*" I asked. "W-e-l-l," Ed drawled, "That's the way it starts." In a few minutes, the puddle was quite substantial and I asked Ed to "dig her in again" to get the best possible well.

The hole needed to set Ed's machine down 8' was probably four times the volume of the previous hole, with a ramp fully twice as long. Now we were encroaching upon territory earmarked for the house. Had we approached the hole from the south instead, we would not have disturbed the soil beneath the house site, and I could have built at the desired location—maybe—close to the well and tucked into the bank. As it was, we had to resite the house 30' east of the original location, which was not a total disaster on our large lot, but would have been on a small lot.

The well was built of 24"-diameter concrete culverts, three 8' sections and a 5-footer. (*See* 9-1.) The 5' section was set on about eighteen inches of 3"

washed stone. Backfill was more of the washed stone to a depth of 5'. I covered the washed stone with 6-mil black polyethylene to keep the stone clean. Further culvert sections were backfilled with the original claylike material which came out of the hole. The well was built just prior to our trip to Scotland in December, 1980. After the spring thaw, we were pleased to find 13' of water in the well. In December, 1982, during another drought, the well maintained a 4' depth with excellent return during pumping.

Aside from digging in from the right direction, I would make one other change if I were doing the well again. I would cement the culverts together. All water would enter the well from the bottom that way, as water seeks its own level. As it is, some red silt enters the well through the small spaces where

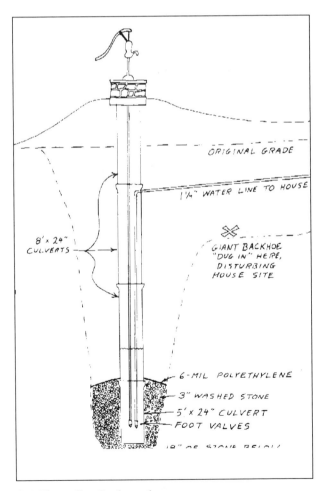

9-1. The well at Earthwood.

the culverts join. After two years, I had to clean the silt out with an electric mud pump, a difficult and time-consuming job. Fortunately, the soil has now stabilized.

WAITING FOR DRY GROUND

We knew that Earthwood was a big project and wanted to make the earliest possible start in the spring. Dennis Lee, who had worked with us on the Log End Cave project, flew into Montreal from Florida with his family on March 20. 1 picked them up at the airport. The snowstorm which greeted them reminded Dennis of the realities of the harsh—and long—North Country winters. The Lees moved into Log End Cottage, a half mile from the Earthwood site. Dennis became my right-hand man on the project, returning to the Florida sunshine in November.

The "site," as the gravel pit became known from the first, was very wet indeed, as snow was still melting. The clay layer which the roadbuilders had struck at about the 5' depth did not allow rapid percolation of ground water. Conditions were clearly two or three weeks away from ideal for building our sand pad, upon which the main building would float. But there was plenty to do in the meantime. We built all our window frames, so that they could get a few months of seasoning prior to laying them up in the cordwood walls. (The building of the frames is described in the cordwood masonry discussion.) We hauled cordwood from Mooers Forks, twenty miles away. Most of our cordwood was cut from old split-cedar fence rails. I paid $25 per 16" face cord ($75 per full cord) for the material, but had to haul it myself. The wood was light, so we stacked it high on our little Toyota pickup truck. Several round-trip journeys were necessary.

We removed the twenty thermopane units from their temporary position in a greenhouse at Log End and stored them away carefully for installation at Earthwood in the autumn. We cut a second driveway through the woods and brush to the west of the site, in order to make a shorter and more pleasing entrance from the road. We collected and sorted stones for the masonry stove, good facing stones in one pile, clean "rubble" in another. All of this time-consuming work was useful and necessary.

On April 29, 1981, we literally "broke ground."

FROST

Northern New York can reasonably expect frost depths of four feet, although an even greater penetration is possible. The builder must be alert to the possibility of frost heaving. When water freezes, it expands 4 percent. This is why frozen pipes burst and the beer bottle which we put in the freezer for quick chilling—and forget about—breaks. There is nothing we can do to resist this powerful force. By the traditional method in our area of burying footings to a depth of 4' the builder is attempting to outguess the frost depth. This usually works. The

9-2. An alternate method of protecting a building against frost heaving, as per Log-End Cave. Frost penetrates at a 45° angle.

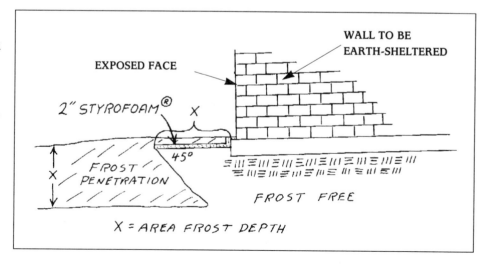

EXPOSED FACE

WALL TO BE EARTH-SHELTERED

2" STYROFOAM® X

45°

FROST PENETRATION

X

FROST FREE

X = AREA FROST DEPTH

building is safe as long as frost depths do not significantly exceed 4'. But in the unusual year when a soaking November rain is followed by a long, hard freeze without benefit of a snow cover for insulation, well, the best guess may not be quite good enough. And, too, setting footings down 4' is long, troublesome, expensive work. In the case of a basement foundation or an underground house, this is not a problem, unless there is some portion of the earth-sheltered storey which is less than 4' deep. In that case, the choices are: (1) attempt to outguess the frost with a standard frostwall footing of a depth commonly used in your area; (2) use Dow Styrofoam® or an equivalent product to protect against frost penetration as shown in 9-2; or (3) "float" the building on a percolating pad, as described below.

THE FLOATING SLAB

Frost heaving is caused by wet earth freezing and expanding, nothing else. Choices (1) and (2) above rely on successfully guessing the maximum frost penetration. Choice (3), the floating slab, gets to the heart of the problem by carrying the water away before it has a chance to freeze. The procedure is as follows:

First, all organic material, such as sod and topsoil, is scraped from the site, preferably to a tidy pile nearby. This should be done carefully, so that this valuable resource can be used later on the roof, in the garden, or as topping on the earth berms. Typically, the undisturbed subsoil will be found at a depth of 6" to 12", although this can vary. A pad is built up on the undisturbed earth so that its final grade after compaction is at least a foot above the surrounding finished grade. The pad is built of some good percolating material such as coarse sand, gravel, or crushed stone. If sand or gravel is used, the material should be compacted in runs of 6" depth at a time. Sand or sandy gravel should be compacted wet, using a power compactor hired at tool rental stores. Pounding dry sand is an exercise in futility.

There are two different styles of hand-controlled power compactors. The older style, known unaffectionately as a kangaroo, actually jumps up and down, and takes the operator along for the ride. The work is about as easy and relaxing as operating a jackhammer. The newer style is a heavy machine with a curved-up leading edge. A weight inside the machine pounds on the heavy plate, but the operator is not involved in the actual pounding. His only job is to steer the thing, infinitely easier than controlling the kangaroo.

The pad thus created might typically have a compacted depth of 18" to 24", depending on how much topsoil had to be scraped away. I extend my pads 3' or 4' beyond the edge of the footing location. A round building with footings 39' in diameter will be floated on a pad of 45'–47' in diameter, for example.

No. 2 crushed stone (1"–1½" diameter stones) makes an excellent pad. Railroad ties and tracks are supported on a berm of this material for frost protection, and a house requires much less of a foundation than a 200-ton locomotive travelling at 80 mph. The disadvantage of the crushed stone pad is that it is difficult to dig for the setting of footing forms. The easiest way to do this is to set the forms on the second or third run of crushed stone, and then to spread the last run carefully by hand.

THE BEST LAID SCHEMES O' MICE AND MEN GANG AFT A-GLEY [5]

Our intent was to sculpt into the banking at the back of the gravel pit, creating a small amphitheatre for setting the house part way into. The material which would come out of the amphitheatre would be used in berming the rest of the northern hemisphere later, cutting down on backfilling. Unfortunately, almost as soon as we started cutting into the 5'-high banking, we encountered water… right on a line with Bill Rivers's "vein," which he had described with his witching rods. (I have become a complete believer in water witching, although I have no idea how it works. Over a year later, a skilled water witch attending one of our workshops found the same vein using a forked hardwood branch.)

We could not build a 240-ton house over a spring; there would be just too many complications, including hydrostatic pressure and trying to keep the water out of the house. Reluctantly, we resited Earthwood yet again, this time 20' forward (south) from the bank. This was worse than the previous move east from the well, as the advantage of the two levels formed by the edge of the gravel pit was largely lost. Months later, this second resiting of the house cost us several hundred dollars in extra earth-moving.

A PAD OF SAND

We built the pad at Earthwood out of sand, as it was indigenous to the lot. There were veins of sand in the middle of the gravel pit, just 100' or so from the pad location. Ed Garrow pushed the sand to the site with his bulldozer. The first few pushes did not accomplish much, as most of the sand slipped to the sides of the blades, but the experienced operator was not deterred. After a few passes, Ed was pushing the sand between the two dikes he had created, spilling very little indeed.

We set depth stakes with a surveyor's level, for Ed's guidance. A red flag tied near the top of the stakes showed him the desired grade. Dennis and I stood by with shovels to help, but Ed's experience paid off, and very little hand leveling was necessary. When a 6"-deep run was finished, Dennis and I watered and compacted the sand, while Ed gathered more sand together at the other end of his trench, or spread gravel on the new driveway. After compacting in both directions and a third time in a spiral direction (9-3), we would call Ed for more sand (9-4). At

9-4. Additional runs of sand are built up and compacted, to a final depth of at least 18".

9-5. The finished pad. The extension near the truck is for the greenhouse.

9-3. Power compacting the sand pad.

the end of the second day, we had a well-compacted, almost flat pad about 18" in depth. We finished levelling the pad by hand. The finished pad, including the extension for the greenhouse, is shown in 9-5.

If good percolating material for building a pad is already on site, use it. Where drainage is good, frost heaving is rarely a problem anyway. If material must be brought in, calculate the volume and add 20 percent for compaction in the case of sand or sandy gravel. Crushed stone will not compact. In order of preference, crushed stone is best (although not the easiest to work with), followed by coarse sand, then gravel. Remember when comparing prices that haulage is very often more expensive than the material itself. Both sand and crushed stone sell for only $6 to $7 a ton in our area (1991). Had we needed to bring percolating material to the site, we would have required about 90 cubic yards, approximately 15 normal dump-truck loads. The cost would have been $700 to $800 and we still would have wanted a bulldozer on hand to assist in the spreading. The primary rule for building an affordable home is: *Make good use of what you've got on site.*

10
The Footings

The footings and floor can be poured together—a *monolithic* slab—or they can be poured separately. My preference on a large building is to pour the footings one day, and the floor later, using the footings as a guide for screeding the floor. With smaller slabs, such as those for the sauna and shed, it is more convenient to pour monolithically. There are extensive preparations for both footings and floor, and there is no real economic advantage in pouring both at once as far as ordering concrete is concerned; at least a full truckload of concrete will be needed for each pour, so there will be no "small load" charges. It is easier for the inexperienced to pour on separate days, although professionals would rather finish the whole job in one day. We had to have concrete trucks on two different days anyway, because the buttresses—discussed in this chapter—had to be poured after the footings.

PREPARATION FOR THE FOOTINGS POUR

All measurements in a round house are based on the center point, so it is imperative that this point be firmly fixed until the walls begin to rise. We drove a strong two-by-four stake into the compacted sand at the center of the pad. A temporary stake had been used to determine the radius of the pad itself, so the new one was a replacement. A 16-penny nail was driven into the top of the stake, a half inch sticking out for fastening the end of our 50' tape. It is a good idea to tie a bright red band around this and every other important stake on site, in order to minimize the chance of their being tripped on or kicked.

ORIENTATION

Of equal importance with the center point, there are eight other external reference points which should be established early in construction, the primary compass points: north, northeast, east, southeast, south, southwest, west, and northwest. These points will be the only means of reference for determining room and rafter locations. Unlike a rectangle, a circle has no unique reference points except the center. Although it is not critical that the points are exactly attuned to true compass directions, it is imperative that they be correct in relationship to each other. Having had a long fascination with such ancient astronomical sites as Stonehenge and Callanish in the British Isles, I wanted to make sure that Earthwood was exactly oriented on a north-south axis. My reasons were romantic and spiritual as much as anything else, although I can't help but delight in the idea of future archeologists trying to work out the intent and purpose of the building. To accomplish this precision of orientation, I had little choice but to beg the help of my good friend—and surveyor of meticulous accuracy—George Barber.

After a fascinating evening of beer and blather, during which he explained several different means of obtaining north and south using stars and the sun, George asked me, "How accurate do you want to be?"

"Can you get us to within a half degree?" I asked.

George laughed. "I can get you to within half a minute."

The next day, George and his son-in-law, Peter

10-1. George Barber, kneeling, uses his transit to lay out the eight directional stakes with incredible accuracy.

Allen, came to the site, and, using a wonderful old transit and an accurate magnetic adjustment, laid out the eight compass points. (*See* 10-1). We drove sturdy stakes at the appropriate locations, each about a yard outside of the perimeter footings. The exact points were marked with small finish nails, and George carefully red-tagged the stakes for safety. The eight points had been laid out with only the transit, but when we checked their relationship to each other using a tape, we found that six of the measurements were exactly right, and the other two were 1/16" off. I was quite willing to believe an accuracy of half a minute.

I relate the story above for interest's sake. The reader who is unimpressed by such things need not concern himself with the exact orientation. What he *should* do, however, is to approximate north or south using a compass, the sun,* or the North Star, and

* To use the sun to obtain solar south, read the morning paper to get the times of the sunrise and sunset. The sun will be in solar south halfway between these times.

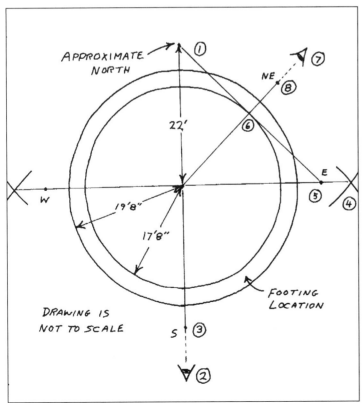

10-2. 1. Drive north stake 22' from center. 2. Extend the line from the north stake through the center stake and on in a southerly direction. This line can be eyeballed quite accurately. 3. Drive south stake 22' from center. 4. To find the east and west points: Clip one end of a 50' tape to the south-point nail, and describe an arc in the sand near the eastern location, using a radius of about 1½ times the center-to-south distance, or 33' in this example. Do the same near the western location. Now clip the tape to the north nail and draw two more 33' arcs in the sand, intersecting those already drawn. The intersections are exactly east and west of the center. 5. Clip your tape to the center nail and have someone hold the tape so that it passes exactly over the east intersection. Drive a stake 22' from center and mark the east point with a nail. Do the same for west. 6. To find northeast, southeast, southwest, and northwest: Measure the distance from north to east. Mark the ground accurately at a point halfway between. 7. Extend your tape from the center point through this ground mark and straight on beyond. 8. At 22', drive the northeast stake. The other points are found the same way.

then drive a stake at some given distance from the center, 22', for example, in the case of Earthwood. The other cardinal points can be found as described in 10-2.

FOOTING DIMENSIONS

We chose a 24" wide by 9" deep concrete footing to support our 16" thick wall and its share of the roof load. The formula for determining footing dimensions on conventional block walls says that the width of the footing should be twice as wide as the thickness of the wall and that the height of the footing should be equal to the width of the wall. This is fine for 8" block or poured walls, but does not hold up for very wide walls. We were overbuilt at Log End Cave, where we built footings 24" wide by 12" deep to support our 12" block walls. Paul Isaacson, with whom I taught earth-sheltered housing seminars for *The Mother Earth News*—and who has much more experience than I in poured concrete earth shelters—pointed out that the last 3" of footing depth at the Cave was really unnecessary. So we'd spent 33 percent more for concrete than was needed. A full 9"-deep footing is as much as is needed for a house of even Earthwood's substantial mass. We chose the

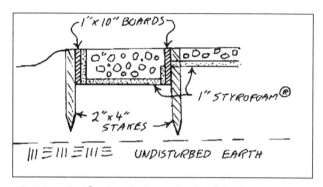

10-4. Forming footings independently of the floor.

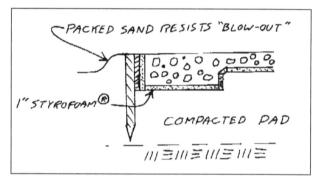

10-5. Forming the footings and the floor at the same time, the monolithic slab.

24" width—and the sizes of the pillar footings and central stove mass footings—to yield a uniform loading on the pad of about 2000 lbs per sq ft. Any settling in the building, therefore, will occur equally.

The placement of the footing forms had to be determined in relationship to the cordwood wall. The inner diameter of the house was to be 36', an 18' radius. The 16"-wide wall would sit in the middle of the 24"-wide footing. The inner radius of the footing form, therefore, would be 17' 8" and the outer radius 19' 8", as indicated in 10-2. We sculpted a round ditch in our pad based on these figures, allowing an extra inch or two for the placement of our one-by-ten forming boards. (*See* 10-3.) We made the ditch about 5" deep and used the sand to broaden the skirting around the edge of the pad. By setting the tops of the footing forms 5" above the mean grade of the pad, we were allowing room for a 4"-thick concrete floor with a 1" rigid foam insulation beneath, as shown in 10-4 and 10-11. After the forms were put in place, some of the excavated sand was

10-3. Dennis Lee sculpts the footing track out of the compacted sand pad.

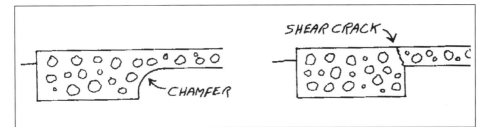

10-6. A chamfered corner where the footing meets the floor will decrease the chance of shear cracking in a monolithic slab.

packed in around the forming boards to resist against a blowout of the concrete during the pour.

The forming boards had to be planed down to almost ⅝" thickness before they would assume the curve. Full-thickness one-by-ten boards were far too stiff. Another option was to cut ⅜" plywood into five 9½" strips for use as forming boards. Eight sheets of plywood would yield enough 9½" wide "boards" to form the footings. An advantage to the monolithic pour is that the inner footing forms are eliminated, as shown in 10-5. In this case, the inner edge of the footing should be chamfered in order to reduce the chance of shear cracking where the footing portion of the slab meets the true floor. (See 10-6.) At the appropriate junctions where two forming boards meet, two-by-four stakes were driven into the sand

pad, with an extra stake between junctions for strength and to keep the midpoint of the boards at the right level.

I have found it easiest to first level the outer forms using a grade level on a tripod, available at tool rental shops. One person holds a calibrated stick on the top of the form board and another reads the grade through the telescope. (See 10-7.) It is imperative that the bubble of the level is centered no matter which direction the scope is aimed in. Keep dogs, kids, and curiosity-seekers away from the level. It is most convenient to have three people on the job, with the third making necessary adjustments to the board height. To lower the forms slightly, tap the stakes with a heavy hammer. *Do not hit the forming boards themselves.* The forms can be raised with a

10-7. Checking the grade of the footing forms with a level and grade stick.

10-8. Dennis levels the inner ring of the footing forms with the outer ring.

pry bar and a fulcrum, such as a brick or a stone. After raising forms, tamp the sand or gravel near the stakes, so that the forms do not settle back into the deep holes.

After the outer forms are set to level—and double-checked—the inner forms can be set to the outer forms using an ordinary 4' level. (*See* 10-8) This tool will be needed to build the masonry walls anyway, so you might as well buy a good one... and take good care of it. We left a full 24" between the inner and outer forms.

INSULATING THE FOOTINGS

A common location for "energy nosebleed" in an earth-sheltered house is at the footings. Warm, moist air inside the house meets the cold footings and that portion of the wall near the footings—resulting in a dew point occurring and condensation forming. We observed this at Log End Cave to a small degree, especially during early summer, when the earth was still cool. The problem can be eliminated by the use of insulation around the footings, as we found at Earthwood. I know of only one product suitable for

use under footings: Dow Blue Styrofoam®. Only Blue Styrofoam® has the necessary resistance against compression for use under footings. The product can be identified by its blue color and the words *Dow* and *Styrofoam*® printed on each piece. *Styrofoam*® will support 5600 lbs per sq ft with only 10 percent deflection. Other polystyrene products, such as beadboard (expanded polystyrene) will suffer unacceptable crushing under the heavy load of the footings, although such products may be perfectly acceptable under a well-drained concrete floor.

I discovered that a 2' × 8' sheet of 1" Styrofoam® could be made to fit almost perfectly between our forming boards by making three cuts as shown in 10-9. The pieces are always laid with the 25" side to the outside, as shown in 10-10. The gaps between pieces are almost imperceptible. No matter what shape of space is to be covered with rigid foam insulation, it is worthwhile to take the time to figure out an effective means of cutting the sheets in order to minimize cuts, waste, and heat loss.

A 9" strip of inch-thick Styrofoam® was placed against the outer form and a 5" strip set against the inner form, as seen in 10-4 and 10-11. These pieces can be fastened to the forms with roofing nails. The idea is to completely wrap the footings and floor

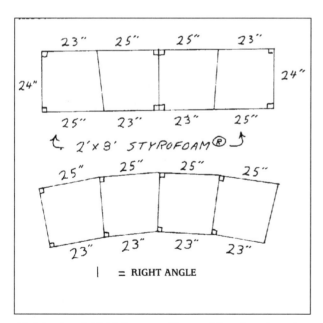

10-9 (top) and 10-10 (bottom). Pieces of Styrofoam® cut in this fashion fit almost perfectly in the footing tracks at Earthwood.

10-11. The wall footings are ready to pour.

with insulation. In the midsection of the United States—Virginia, Kentucky, southern Missouri, and on through the Four Corners region—underfloor insulation can be limited to about 24" in from the footings. In the Deep South or any similar area where cooling is the more common energy problem, the rigid foam under the floors and footings can be left out entirely, as the idea is to dissipate heat away from the house. The earth will not get cold enough to cause condensation problems.

REINFORCING BAR

Rebar must be placed in the footing tracks prior to pouring. We used old ⅝" silo hoops, recycled from a silo which we bought after it had been blown down by a windstorm. These old hoops are ideal, as they were designed for great tensile strength in the first place, and the rust-pitted surface gives a good "grab" to the concrete. If you are buying rebar, use ½" stock (No. 4). If you are troubled by strict building codes in your area, check with the building inspector. Some areas may require three runs of rebar around your footings—although we used two at Earthwood—or insist upon ⅝" (No. 5) rebar.

It is important that the rebar be kept in the bottom half of the pour. The purpose of the reinforcing is to impart tensile strength to the footings, especially to resist settling. Only by placing the rebar in the bottom half of the pour will this tensile strength be gained. I used old broken bricks as supports for our rebar, which kept the material about 3" off the bottom of the track. Keep all rebar at least 2½" in from any edge of the pour. In addition, we used an 18" tie piece of ½" rebar about every 4' around the footing, seen in 10-11. These pieces keep the silo hoops in place during the pour, but also lend tensile strength to the breadth of the footing.

THE BUTTRESSES

In normal construction, we would be ready to pour at this point, but Earthwood is not quite normal and I must digress for a bit to speak of such things as buttresses and arches.

The Earthwood design calls for a heavy earth berm on the northern hemisphere. This berm places a great lateral pressure on the northern part of the house. A good analogy is that Earthwood is like a stone or brick arch lying on its side. Arches are tremendously strong, because masonry materials are phenomenally strong on compression. Professor J. E. Gordon, in his excellent book, *Structures: Why Things Don't Fall Down*, says, "Elementary arithmetic shows that a tower with parallel walls could have been built to a height of 7000 ft or 2 kilometres before the bricks at the bottom would be crushed."[6] Figure 10-12 shows how an arch works. The vertical loads on the arch are transferred to lateral loads by the *voussoirs*, the innermost masonry blocks, usually tapered. The *abutments* of the arch must be strong enough to supply a reaction to the accumulated load of the voussoirs. The arch cannot fall inwards as long as strong abutments are used, because the underside of the arch is in compression and we do not even begin to approach the crushing strength of the masonry.

At Earthwood, the voussoirs are 8" concrete corner blocks laid widthwise in the wall like log ends. The abutments are strong concrete buttresses, designed to supply a reaction force to the lateral pressure on the northern hemisphere of the building, as shown in 10-13. These buttresses must be strong enough to prevent the building from moving suddenly to the south.

To accomplish this, I designed the buttress system illustrated in 10-14 and 10-15.

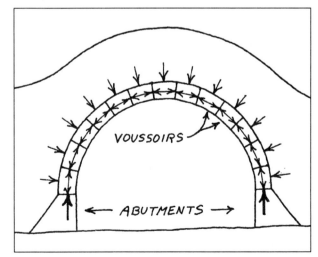

10-12. An arch. Arrows indicate the lines of thrust. The large arrows at the abutments indicate the reaction thrust.

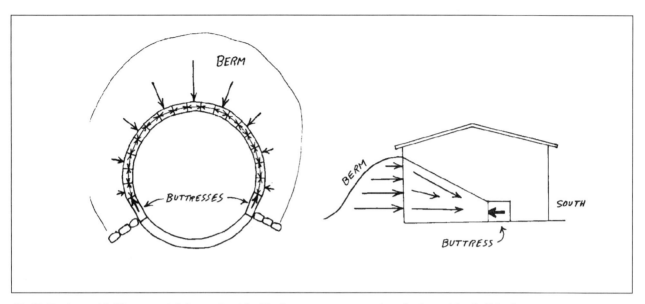

10-13. Earthwood is like an arch lying on its side. The buttresses react against the lateral load of the berm.

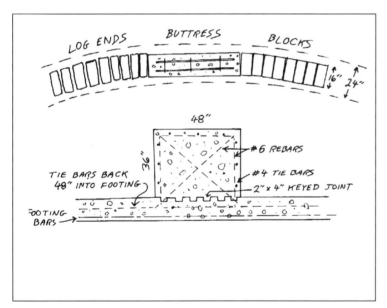

10-14. (left) Buttress detail. 10-15. (right) The buttress cage is fabricated from No. 6 reinforcing bar.

First, I went to the local foundry and had the necessary rebar pieces made, all out of ¾" (No. 6) reinforcing rod. (*See* 10-16.) The square pieces were heat-bent and welded to close the square. The diagonal pieces are bent at a 45° angle and are tied back 4' into the footings in each direction. The entire buttress cage assembly is tied to the footing reinforcing bar with forming wire. We did not use the two-by-four keyway detail shown in 10-15, but it is a good idea suggested by architect Marc Camens. Clearly, the buttress skeleton is well and truly fastened to the footing. After the footings were poured, a forming box was built around the buttress skeleton, to be filled later during the floor pour.

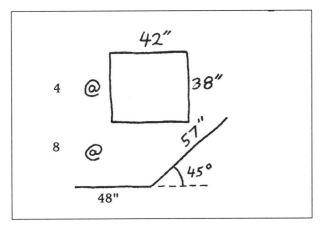

10-16. The pieces required to build two buttress cages. Use ¾" (No. 6) rebar.

STONE-MASS FOOTING

The footing for the large cylindrical stone mass at the center of the house is poured at the same time as the perimeter footing. We chose a diameter of 7' for this footing, in order that the load would be similar to that of the perimeter footing, around 1600 lbs per sq ft. The pillar footings are also designed for a similar load. The point of these calculations was to assure that the house would settle equally on the pad. Again, a depth of 10" was chosen, with an inch of Dow Styrofoam® used below the concrete, so that the true depth of concrete in all the footings is actually 9".

10-17. The central-mass footing is ready to pour. The center stake is left in the concrete.

This is ample, although I would not recommend a footing depth as low as 8" for a house of this mass. (All told, Earthwood with a fully saturated roof weighs about 240 tons.) Rebar was crisscrossed throughout the stone mass footing, as can be seen in 10-17. Because of the rather tight circumference of this footing, it was necessary to form the pour using ¼" plywood, which has good strength and flexibility.

THE PILLAR FOOTINGS

There are seven pillars at Earthwood, one point in the octagon having been eliminated to accommodate a billiard table. The northeast and northwest pillar footings, therefore, were designed a bit larger than the others to yield equal compression forces on the pad. The five smaller pillar footings have diameters of 3', while the other two have diameters of 3' 8". These pillar footings were not formed, but merely sculpted out of the compacted pad. A disc of Styrofoam® was placed at the bottom of each 6"-deep hole and the sides of the holes chamfered and wet-packed just prior to the pour. Only the bottom half of the pillar footings were filled when the perimeter footings were poured. The concrete was left rough so that the floor pour would bond well to the cold concrete later on. This was done for safety and convenience. If the pillar footings are completely poured along with the floor, there would be seven 4"-deep holes for people to stumble into. And there would be a greater chance of breaking loose sand into the holes during the pour. Rebar was crisscrossed through the pillar footings, in similar fashion to the stone-mass footings.

CONCRETE CALCULATIONS

I do not believe in mixing my own concrete. There is very little saving in material cost, unless the site is a long way from a concrete batch plant, and there will be a number of weak "cold joints" where one pour meets the next. Ready-mixed concrete is quick, easy, and of more consistent quality. One of the important jobs, therefore, is to make an accurate calculation of the amount of concrete required. This is probably as good a time as any to go over some of the basic formulas for dealing with circles and cylinders.

The area of a circle is found by multiplying the constant pi (π) by the radius squared. $A = \pi r^2$. I use a value for pi of 3.1416. For example, the area of a slab with a diameter of 40' (radius is 20') would be: $3.1416 \times 20' \times 20' = 1256.64$ sq ft.

The circumference (perimeter) of a circle is equal to the diameter times pi. $C = \pi d$. Or: $C = 2\pi r$. Example: a circle with a diameter of 10' has a circumference of 31.416'.

The volume of a cylinder is simply the area of the circular base times the height of the cylinder. $V_{cyl} = \pi r^2 h$. If the 40'-diameter slab described in the example above had a uniform thickness of 6" (0.5'), the volume would be: 1256.64 sq ft $\times$ 0.5' = 628.32 cubic feet. Concrete is measured by the cubic yard, however, so it is necessary to divide the answer by 27 to give cubic yards: 628.32 cu. ft/27 cu. ft per cu. yd = 23.27 cu. yds.

The Earthwood footings are like a doughnut: a cylinder with a large doughnut hole (smaller cylinder) removed. The large cylinder, with its radius of 19' 8", has a volume of: $3.1416 \times 19.667' \times 19.667' \times 0.75'$ (9" true concrete depth) = 911.35 cu. ft. Dividing by 27 gives 33.75 cu. yds. The doughnut hole which will not be poured has a volume of $3.1416 \times 17.667' \times 17.667' \times 0.75'/27$ cu. ft per cu. yd = 27.24 cu. yds. The difference is 6.51 cu. yds to go in the perimeter footing. The central mass footing is: $3.1416 \times 3.5' \times 3.5' \times 0.75'/27$ = 1.07 cu. yds. Similarly, the two 5"-deep pillar footings with the 1' 10" radius yield a total of 0.325 cu. yds. and the five pillar footings of 1' 6" radius require a total of 0.545 cu. yds The grand total of concrete needed for the first pour, then, is: 6.51 + 1.07 + 0.325 + 0.545 = 8.45 cu. yds Call it 8½ yds. Order 3500-lb ("six-bag") concrete for the footings.

Our local trucks carry about 9 cu. yds, so this worked out quite well, but 8½ is enough for safety, as the Styrofoam® up against the footing forms displaces some concrete, and if we're just a wee bit short for one or two pillar footings, this does not really matter too much, as this can be made up during the floor pour. During the floor pour, allow a greater safety margin, as you will not want to send a truck back for an extra yard. You will be charged a hefty "small load" charge.

It is always a good idea to have some place to dump an extra ¼ to ½ yard of concrete. You are paying for it, so you might as well use it. You might have a walkway framed up for any extra concrete, or a small pad for trash cans or propane gas cylinders. With our small "leftover," we poured a footing around the well for supporting stonework. If you don't use it, you may find a hard grey pile of concrete near your property where the driver cleaned the truck.

THE POUR

Four or five strong bodies are sufficient for the footing pour. Ask the driver for a stiff mix; a 4" to 5" "slump" is good, should he request such a figure. *Slump* refers to how far the concrete will sag or slump when a test cone is removed from fresh concrete. While "soupy" concrete will be easier to spread, the full strength will not be there and the concrete will be much more prone to shrinkage cracking.

Tools required for pulling and moving concrete are strong rakes or hoes and long-handled shovels. Make sure the concrete truck can "chute" the material into any part of the footing. (*See* 10-18.) Clear access to the entire job is imperative. Wheelbarrowing concrete is hard work and time-consuming.

Flattening the top of the footing (called "screeding") can be accomplished with a 4'-long piece of two-by-four. The footing can be smoothed with a flat trowel, but don't be too fussy about this, as a slightly rough surface will provide a better grab for the masonry walls to follow. The concrete is left rough within the confines of the buttress cages, so that there will be a good friction bond at the cold joints where the buttresses meet the footings. (Or, you may choose to employ Marc Camens's "keyed joint" detail shown in 10-15.) If trowelling produces too smooth a surface, the middle 16" of the concrete can be scratched with the corner of a trowel after a couple of hours, to improve the friction bond to the masonry wall.

With a rectilinear design where there will be earth pressures on the walls, I recommend the inclusion of a keyed joint in the pour. The joint is formed by setting a piece of wood flush with the surface of the concrete, halfway between the forms. A good

10-18. Pouring the footings.

keyway will be formed by a two-by-four ripped down the center with a circu.lar saw. If the blade is set at an angle of about 75°, instead of 90°, an excellent draft angle will be formed to facilitate easy removal from the partially set concrete. Oiling the keyway board is also strongly recommended. The resulting keyed joint will look like Figure 10-19.

Note that the draft angle is kept to the outside of the footing. Later, the first course of blocks can be

firmly tied to the footing by filling the cores halfway with mortar or concrete. With a round house, it is not necessary to include a keyed joint or to brace the first course of blocks in any way. The inner surface of the wall of a round house is on compression, as was discussed.

Footing forms can be removed after a day or two, depending on drying conditions. You can make footing forms easy to remove by painting them with oil prior to the pour, although this is not necessary where Styrofoam® already protects the boards from contact with the concrete. Be careful not to spill excessive oil on the Styrofoam®, as this causes a rapid deterioration.

FORMING THE BUTTRESSES

Although we used 12"-wide buttresses at Earthwood, I would pour them 16" wide if I were doing it again. The idea behind the 12" buttress was to enable us to build a 4"-wide internal facing of bricks or small logends, but we changed our ideas on this when we decided to go with concrete blocks below grade. We faced the inner surface of the buttresses with 4" solid concrete blocks, so we might as well have poured 16" wide right from the start. We used old two-by-six tongue-and-groove silo boards and double-headed forming nails to form the buttresses. With a 16"-wide pour, I would recommend strong bracing against the centers of both sides of the forms, or tying the sides of the forms together with wire right through the concrete, as is done with slip-form stone construction. The wire is cut loose from the forms later and left in the concrete. The lateral pressures exerted on the sides of the form by a 16"-thick wall are phenomenal and a concrete blowout will spoil your whole day.

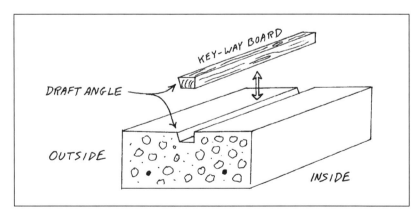

10-19. A keyed joint in the footing.

11
The Floor

There are a tremendous number of jobs that have to be done prior to pouring the floor, so be prepared.

UNDERFLOOR PLUMBING

Opinions are mixed as to the wisdom of putting plumbing under a concrete floor. Charles Wing, writing in *From the Ground Up* (Little, Brown, 1976), says:

> The slab disappointed (those) who thought they could neatly dispose of all that ugly plumbing and wiring by pouring concrete over it. Unfortunately, pouring concrete over anything doesn't leave much flexibility... They found that when the pipes burst due to differential expansion and contraction and just plain corrosion, the appropriate tool was the jackhammer instead of the pipe wrench.[7]

You have been warned.

However, I continue to place outgoing plumbing under the slab and am by no means alone on this. Mr. Wing's advice is well taken, and emphasizes the need for meticulously careful work under the slab, so that problems will not occur. This is one area where I would advise consulting with an installation plumber who has a reputation for quality work. It may be worthwhile to hire the plumber to do the work.

We did our own underfloor plumbing at Log End Cave and it has not caused us any trouble in five years. Nevertheless, I was still extremely cautious at Earthwood, and called in for consultation a plumber friend of mine, Bob Evans, who, with his wife, Aggie, was building a 45'-diameter cordwood house on a monolithic slab about ten miles from us.

"What pipe are you going to use under your slab, Bob?" I asked.

"Cast iron!" came the reply like a shot. "Yeah, it's more expensive than plastic, but you'll never have to worry about it."

I figured that if Bob, a plumber of long experience, was confident with cast iron under his floor, this is what I would use, too. In fairness to another viewpoint, others in the contracting business have said that the expense of cast iron was unnecessary, that excellent strong plastic pipes are on the market today which are just as good. (A few years later, we did the underground plumbing at Mushwood with ABS plastic, which was cheaper, *much* easier and will, I'm now sure, last at least as long as cast iron, probably longer.)

An alternative to the underfloor plumbing concept—and of particular interest to those who are adamant about a wooden floor anyway—is to set concrete blocks on the slab at frequent intervals for the support of a light floor-joist system. Thus, plumbing and electrical systems can both be reached by raising floorboards instead of lifting concrete. Mr. Wing's commentary on this in *From the Ground Up* is worth reading.

At Earthwood, our septic tank serves both the toilets and the downstairs bathroom. The upstairs bathtub, washbasin, and kitchen sink go to a soakaway—also called a dry well—located about 30' southeast of the house. This is called a grey water system. The sauna and hot tub also drain into this soakaway, as will a washing machine if we ever have one. The reasoning behind this is to keep the chlorine or bromine of the hot tub out of the septic tank, where it can kill the digesting bacteria. A washing machine or sink where bleach or certain soap pow-

11-1. Underfloor plumbing. The cylindrical cap is set at the same grade as the top of the footings.

ders might be used should not be directed to the septic tank either, although many are.

A soakaway is a large pit, 6'–8' in diameter and an honest 6' deep. (In soils with excellent percolation, such as coarse gravel, these dimensions can be scaled down somewhat.) The hole is filled with stones and covered with some sort of filtration barrier prior to final grading. Straw or hay makes a good filtration mat, as does builders' felt paper or discarded cement bags. In the north, it is wise to cover the soakaway and the pipe going to it with an inch or two of polystyrene insulation to prevent freezing.

It is a good idea to plan the plumbing so that all traps are accessible. If a bathtub or shower is to be used on the first floor, then some creative design is necessary to prevent the traps from being located under the slab. One option is to elevate the bathtub, but this could be frowned upon for safety reasons and might not impress some future buyer of the home. The other option is to "box out" an area of the floor—an 18" square will be sufficient—under the projected bath drain. This stops concrete from being poured in this area. In case of trouble, the trap can be dug out of the sand pad.

All outflow plumbing must be set firmly in the compacted material of the pad, so that no settling can occur. The proper slope for all underfloor pipes in the waste system is between ⅛" and ¼" per ft. I try to keep halfway between these figures, about ³⁄₁₆". Using a 2' level, then, the slope will drop ⅜" over the

length of the level; with a 4' level, the drop should be about ¾" along its length.

Find the exact location of all of your plumbing appliances. This is why it is imperative to have your eight stakes corresponding to the eight major compass points. We found it useful to put an 8"-square piece of wood at the exact location of each of the future eight-by-eight posts. These posts will fall at the approximate center of each of the pillar footings, but measure exactly off of a line connecting the center to the appropriate compass point stake. With the mock "posts" in place, it is an easy matter to draw all the pertinent internal walls in the damp sand with a stick. Using the floor plan, the exact locations of all waste pipes can be determined. Get a good book on plumbing or seek advice from a plumber or the plumbing supply house, unless you have decided to hire a professional for the installation. With plumbing and electrical systems, hiring a contractor may be a good decision, but shop around. Get estimates from at least three people, and pay attention to reputation, especially with regard to follow-up work.

Leave all waste pipes sticking up a foot or so and wire them firmly to a driven stake. Very carefully cap or cover the tops of all pipes so that concrete or other flotsam and jetsam does not find its way into your waste pipes. There is nothing more attractive to a toddler than pouring sand down a tempting pipe. That just has to be what it's there for! With a toilet receptacle, which must be set at the exact floor level,

11-2. Checking the slope of the underfloor drains.

cover the flange with a "Red Cap" cover made for the purpose, as seen in 11-1. The device is exactly what the name implies: a red plastic cover cap set level with the floor. It can be cut away later with a knife when it is time to install the toilet.

Do not forget to connect the 3" vent stack to the main line as required by code, and without which your plumbing will work sluggishly. The vent allows air to return to the pipes as water rushes out. It serves much the same purpose as does punching two holes instead of one in a quart can of oil.

The main waste line beneath the floor will be 3" pipe, the same as comes out of a toilet receptacle. Branch lines from sinks and bathtubs can be 1½" pipes; a washing machine requires a 2" pipe. The 3" waste line goes under the footing—this is easy to dig in compacted sand—and expands to a 4" house drain outside of the footing, which carries on to the distribution box and septic tank. In gravel or crushed stone pads, it would be better to install the pipe under the footings prior to the footing pour. Depending on the location of your water source, incoming plumbing may also have components beneath the slab, and these must be considered.

ELECTRICITY

All our electricity is made from wind and solar power and is 12-volt DC. As we are not connected to the power company in any way, we are not subject to their inspections. If you plan on connecting to power company electricity, or plan to build in a coded area, you will need to do all of your wiring "by the book." Bob Evans, building his round cordwood house a few miles away, had to keep this in mind during his slab-pour preparations. Bob chose to run underfloor conduit from the area where the service entrance would be to various internal wall locations. Most of his electric lights and fixtures, then, are located in internal walls. Again, the importance of being able to determine the exact location of internal walls is demonstrated.

My personal preference with cordwood houses is to make use of the newer style of quality exposed wire mold (also called "surface-mounted raceway"), which comes in a variety of colors and is code-approved. Almost all new industrial buildings, as well as many homes and public buildings, are making use of good-

quality exposed wire mold. The advantages are ease of installation, ease of changing or expanding circuits, and ready access in case of trouble. Attractive exposed wire mold is a godsend when working with exposed plank-and-beam roofing. More and more people are breaking out of old patterns and discovering that there is really nothing objectionable about exposing structural systems, plumbing, and electric… and that it makes a great deal of sense in terms of ease and economy of construction, as well as low maintenance ongoing costs. The problem with exposed electric is that surface-mounted wire mold will more than double your wiring cost.

At Earthwood, then, the running of electric wires did not come into play with regard to the slab, although someone with "regular" electricity may wish to run conduits under the floor.

THE HOT TUB INSET

There was one unusual job which we had to attend to at Earthwood prior to the floor pour. During the planning stage, we had decided to include a cordwood masonry hot tub on the lower level. Our intent was to heat the water using a special wood stove having a hot-water jacket surrounding the firebox. Furthermore, we wanted to circulate the water through the stove without using an electric pump, as we would be relying on wind-generated 12-volt power as our primary source of electricity. The hot water made by the stove would have to be delivered to the tub via the thermal siphon principle.

A thermal siphon works on that ever-handy law of physics which forces warm fluids to rise. The greater the differential in height between the heat source (the stove) and the tank (the hot tub), the greater is the power and drive of the thermal siphon, just as the taller the chimney, the greater its draught. In order to get a sufficient height to the thermal siphon, we decided (after consulting with Renaissance man George Barber) to set the stove 16" into the floor. To accomplish this, we excavated about 20" into our sand pad, and poured a 4"-thick concrete slab with dimensions of 48" × 66". We chose these dimensions to allow us to enclose a free space of 32" × 62" with ordinary 8" concrete blocks. This gave us ample space for our stove, which we had already bought. The blockwork was surface-bonded (see

Chapter 13) and wrapped in a double layer of 6-mil black polyethylene and 1" of Dow Styrofoam®. The inset can be seen in 11-8, and in the illustrations in Chapter 23, The Cordwood Hot Tub.

UNDERFLOOR DRAINS

A cheap insurance against a damp floor is to set an underfloor drain into the sand or gravel pad. This is particularly important in an earth-sheltered house. The drain is 4" perforated flexible plastic tubing, complete with a nylon or fibreglass filtration sock designed to keep the little slits in the tubing from clogging with silt. The technique is to bury the tubing so that its top just barely shows through the pad, as seen in 11-3. A slight slope towards the discharge end is advised. All points of the pad should be within 6' or so of the underfloor drains. Figure 11-4 shows a typical configuration for a square building and an approximation of what we actually did at Earthwood. In either case, the outflow (low end of the drain) is taken under the footing at some convenient point and slopes away to a soakaway pit, as with the grey water system. We dug our soakaway about 30' from the south side of the house. Both the underfloor drains and the footing drains terminate at the soakaway, roughly 8' in diameter and about 5' deep. We

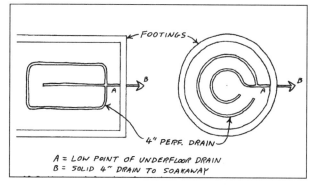

A = LOW POINT OF UNDERFLOOR DRAIN
B = SOLID 4" DRAIN TO SOAKAWAY

11-4. Positioning underfloor drains for good coverage.

filled the pit with leftover stones from the masonry stove, and covered it with a filtration mat to stop excessive silting.

INSULATION

Because of our cold climate—we live near Montreal—we decided to insulate the entire slab with 1" of Styrofoam®. It is a good idea to use a scale plan of the footings and floor to determine the most efficient configuration for laying and cutting the sheets. This is particularly worthwhile with a round structure. Although we used Dow Styrofoam® (extruded polystyrene), a good-quality expanded polystyrene—also called EPS and beadboard—would be acceptable providing underfloor drainage is very good. An inch of Styrofoam® will have a 20-30 percent higher insulative value (R value) than an inch of beadboard (depending on the brand), but the beadboard may be less than half the price. We tried to fit the Styrofoam® as nearly as possible to the footings, without being excessively fastidious. We cut holes in the Styrofoam® corresponding to the pillar footings, which were already insulated beneath their lower halves, previously poured.

11-3. The 4" perforated underfloor drain is set into the sand pad.

REINFORCING

The purpose of wire mesh reinforcing is to hold the concrete together if and when it cracks. The com-

11-5. The Styrofoam® insulation and wire mesh reinforcing is installed. The floor is ready to pour. The two buttresses are formed and ready to pour.

monly used mesh is called "six-by-six by ten-by-ten." This rather cryptic phrase means that the mesh has a 6" × 6" grid system and the wires running in each direction are No. 10 gauge. The roll of mesh is commonly 5' wide and 150' long, yielding 750 sq ft per roll gross. The effective yield is about 10 percent less, due to the recommended 6" overlap between runs. Tie adjacent runs of mesh together with forming wire. Illustration 11-5 shows the floor insulation and wire mesh.

THE DAMP-PROOFING LAYER

There is a twofold purpose to the damp-proofing layer: to stop "rising damp," and, more important, to retard the set of the concrete, thus decreasing the number and size of shrinkage cracks. A layer of plastic will not dam water out of the house and this is not its purpose. Apart from anything else, a plastic sheet will be inordinately punctured by the trampling of work boots on the wire mesh. The sand or gravel pad itself is the first line of defense against dampness; it draws away underfloor water. The 4" perforated drain helps by drawing water away more quickly than a sand pad alone. If water has made its way to the concrete despite these precautions, the problem is something major, like building below the water table, using poor materials for the pad, or not installing the drains correctly. Polyethylene will be of little use.

But plastic's second purpose—arresting the rapid transfer of water from the fresh concrete to the pad—is valid. This prolongs the curing time of the concrete. A closed-cell product like Styrofoam® insulation will accomplish the same thing, however, so there was no point in using a plastic layer at Earthwood. Further south, where underfloor insulation is not used, a layer of 6-mil polyethylene will help retard the set and improve the quality of the concrete.

THE FLOOR POUR

Again, you will need about four or five strong backs for pulling and pushing the concrete. The floor is more difficult to pour than the footings, so it is good to have at least one experienced floor man on your crew. I always hire my contracting friend Jonathan Cross to help at major floor pours, because of his experience and take-charge manner. Also, he knows how to operate a power trowel, a potentially dangerous tool, which you don't want to learn to operate on your own floor. If one of the four rotating trowel blades catches onto a corner of wire mesh sticking up too high in the concrete… well, you can get literally wrapped up in your work before you can turn the machine off.

11-6. Pouring the buttress.

First we pour the two buttresses. (*See* 11-6.) Use a long rod to tamp and vibrate the concrete into the form to eliminate voids, but do not overvibrate, as this can put a tremendous pressure on even very strong forms and a blowout is possible.

Next we poured the small sauna slab. A drain at the center of the sauna was originally set 1" below the outside footing forms. (*See* 11-7.) The idea was that we would screed the sauna from the center to the outside of the forms, giving a slightly conical shape to the floor. This way, water used for washing in the sauna would run out the drain. Unfortunately I did not have the presence of mind to wire the cast-iron drain to the underfloor plumbing and it floated upwards in the concrete. The floor ended up almost perfectly flat and water would not run out the drain. Later on, when the sauna was complete, I poured a new floor over the original one so that the wash water would run out. Do not underestimate the power of fresh concrete!

One member of the crew should be delegated the responsibility of lifting the wire mesh up into the concrete during the pour. The corner of a rake works very nicely for this job. The floor will be 4" thick. The mesh should be between 1"–2" off the insulation, although this is hard to do precisely. Do not rest the mesh right on the insulation where it does no good; and it should not be near the surface where it can get in the way of the trowelling.

We screeded the floor flat by spanning our screed board from the center mass footing to the perimeter footings, a 14' span. We used a 16'-long two-by-six for the job, as seen in 11-8. It is a good idea to nail a strong handle to each end of the screed board, although this is not shown in the picture. Pulling the screed board back and forth through the concrete is a tough two-person job. Other helpers try to keep the work going smoothly by rough-spreading the concrete ahead of the screed board, giving a little extra as needed with a shovel, or drawing excess away with a rake. We poured concrete right up to the existing footings, without using an expansion joint. On floors longer than 40', an expansion joint of ½" homosote board (or rigid foam) is recommended where the floor meets the footings.

After screeding, the slab is floated, a process by which water is pulled up to the surface with a tool called a bull float (a large magnesium or aluminum float trowel), which can be hired at tool rental shops. Hire sufficient extra handle sections so that the slab can be floated without your walking on the fresh concrete.

Trowelling gives the floor its final smooth finish. The floor

11-7. Small slabs, like that of the sauna shown here, are easy to pour *monolithically*.

11-8. The depression is for the hot-tub stove. The screed board is a 16'-1ong two-by-six.

11-9. Jonathan Cross smooths the Earthwood floor with a power trowel.

can either be hand-trowelled or power-trowelled. With a large floor, such as Earthwood's, I recommend the power trowel. With a small shed or outbuilding, hand trowelling will suffice. In either case, experience is the byword, and the job should be delegated to someone who has it. A good smooth floor is infinitely easier to cover later with almost any surface: carpet, linoleum, tile, slate, concrete paint. Illustration 11-9 shows Jonathan Cross operating the power trowel. The floor must be firm enough to walk upon prior to trowelling.

During hot weather, the floor should be flooded frequently with water to help retard the set of the concrete. When any cementitious material (such as concrete, mortar, plaster, or surface-bonding cement) dries too quickly, shrinkage cracks will form.

Estimate the concrete for the floor by using the same formulas already explained. However, it is a good idea to check the depth of your floor prior to making the computations, as an average floor-depth variance of just ½" can make a 2-yd difference in concrete. Using your screed board to show final floor level, check the depth at each of the 24 points shown in 11-10, and calculate the average depth. This is more accurately done before rigid-foam insulation is put down, so a 1" adjustment to the average figure should be made. The average depth of the pour is the figure to use as the height of the cylinder in the calculation. With the sauna slab and the buttresses, we required about 14½ cu yds in theory, so I ordered 15. In reality, we had about ¾ of a cubic yard left over, just the right amount to pour our greenhouse slab, which Dennis and I quickly formed up

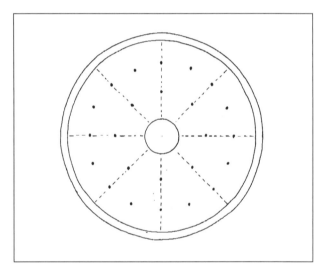

11-10. Check floor depths at these locations prior to the floor pour. Use the average depth as the height of the cylinder in the concrete calculations.

with two-by-four studs on edge. It came out all right, but I should have followed my own advice and had the greenhouse slab nicely formed out and prepared ahead of time. The Styrofoam® and wire mesh were rather hurriedly applied, although the building has not suffered for it.

Prior to wall construction, the locations of the main posts in the post-and-beam framework should be accurately and indelibly marked on the slab. Paint is best. The eight primary compass point stakes are necessary to this job and they will be out of sight once the walls are up a way.

12

Cordwood at Earthwood

This chapter will be short, but not particularly sweet. It is hoped that recounting our experience with very dry hardwood expansion will prevent someone else from having the same kind of problem. Listen:

A PROBLEM WITH VERY DRY HARDWOOD

Our original intent was to build all the external walls, including those below grade, of cordwood masonry. Above grade, we would use cedar for its superior insulative value. Below grade, we would use dense hardwoods for greater thermal mass, as there would be additional Styrofoam® insulation to the outside of the waterproofing membrane anyway. Our mixed hardwood log-ends were cut to length, split, stripped of bark, and properly stacked and covered for about three years prior to use. They were truly "popcorn-dry." When you cracked two of them together, they had the ring of hockey sticks battling for a loose puck.

We attached one end of our 50' tape to the nail at the center stake, and, with a black crayon, described two circles on the footings, one with a radius of 18' and the other with a radius of 19' 4". These circles established the position and thickness of the external walls. We built curved forms of two-by-fours and some of the one-by-ten forming boards from the footing pour. (*See* 12-1.) The forms were built to conform to the outer circumference. They were approximately 10' long and 7' 6" high. We would position the forms and check their plumb each day prior to laying up wood. Laying up cordwood to the forms

12-1. This curved form works very well where a fine pointing job is not essential.

enabled us to keep the outside of the wall relatively smooth, so that we could apply a parging coat of plaster fairly easily prior to the application of the waterproofing membrane. This worked very well, although we could only use the forms on the northern hemisphere of the building, which was to be earth-bermed, as the vertical studs of the form got in the way of doing a pleasing job of the pointing. As the bermed portion of the wall was flush-pointed anyway (as opposed to recessed), without concern for appearance, this didn't matter. On the south side of the building—between the buttresses—we used old cedar log-ends exclusively and kept the wall going up straight with a 4' level.

Log-ends not within one-half inch of 16" were rejected as out of tolerance. On the south side of the house, log-ends were centered in the wall so that there was rarely more than ¼" variation in the protrusion of the wood away from the mortar. On the north wall, the inner surface sometimes had a full inch of relief, as we always laid up to the forms to keep our exterior plastering job easy later on.

Cordwood work commenced on June 10, 1981, and we made great progress, with Dennis Lee mixing mud, and Jaki and I laying wood and pointing. We were finishing 30 to 36 sq ft of wall per day, and I don't think I've seen a more attractive variety of colors, shapes, and textures in a cordwood wall. Maple, ash, beech, and red cherry—split and unsplit—played in harmony together.

It rained on the night of June 21, but we were careful to cover our work each night prior to leaving the site, so I anticipated no problem. On the 22nd, we discovered a small crack between the first mortar course and the first course of wood, but only on the hardwood wall, not where the super-dry cedar was used on the southern part of the circle. The crack was wider on the inside of the building than the outside. On the 23rd, the crack was wider and we could see that the north hemisphere wall was beginning to tilt outward. By the 25th, we joined the two buttresses with a wall 2' high all around, although some parts of the wall were as high as 6'. But, by now the crack had opened to a ½" gap. Other stress cracks were forming in the wall as a result of the large crack, and the wall was 3" out of plumb at 6' high. The line of least resistance was for the wall to tilt outwards, and this it did, hinging up on the base of the outer mortar joint. These conditions were unacceptable for continuing, as we still had another storey

and a heavy earth roof to support. If the downward lines of thrust of the building were to wander out of the middle third of the wall thickness, we would have an unstable and dangerous situation on our hands. With great reluctance, we tore the beautiful hardwood wall down—18 days of work down the drain.

But, things got worse.

Prior to dismantling the wall, I had a structural engineer examine the situation. We speculated on possible causes. There was no problem with the southern portion of the wall made of cedar. We speculated that the sun's heat reflecting off the white slab was causing heat expansion in the masonry, so we tried expansion joints of ½" homosote board every 20' around the back side. We tried a different mortar mix that would harden faster, thinking that outward pressure on the curved wall as the still-plastic mortar was setting might be causing the wall to lean out. We tried using a solid mortar joint on the first course, leaving out the insulation cavity. Nothing worked. The ubiquitous crack kept returning. Great discouragement and even depression was setting in. *Who said building your own house is easy?* In July, embarrassingly, we were conducting cordwood masonry workshops, with still no clear idea of the nature of the devil we were fighting.

Finally, during a conversation with George Barber, something twinked. George was explaining how the old-time stone quarriers would break off a face of granite. In the days when labor was cheap and dynamite expensive, the quarriers would drill holes down into the rock parallel to the face. They would jam the holes with tight-fitting dry hardwood dowels. They watered the dowels and, presto, the wood fibres would swell and break off the stone face. If granite could not resist this tremendous expansion pressure, what chance did our poor walls have? Rainwater collecting on the slab would dampen the first course of the very dry hardwood, causing it to swell. Because of the nature of a circle, already discussed with regard to arches, the inner surface of the bottom course of log-ends becomes very tight on compression. Something has to give. The outside of the wall is on tension, supplying the relief valve. The wall can do nothing but tilt out. The cedar cordwood was so light and airy that the wood fibres could expand within the wood itself without putting a pressure on the mortar joints.

Now we knew the problem: the hardwood was too dry. I had seen other cordwood structures built of hardwood which did not suffer from the same problem, but the wood used was relatively green. Identifying the problem was one thing, while formulating a solution was another. All told, we had lost about 30 workdays with building walls and tearing them down again, not to mention close to $2000 in labor and cement. Jaki and I decided that we could not risk trying another solution, such as preswelling the wood by soaking prior to laying up the first course. We had tried that to some extent with little success anyway. And we didn't have enough cedar to build the entire house out of this gentler wood. Looking at the rapid advance of the calendar and the diminishing balance in our checkbook convinced us to abandon the hardwood wall in favor of a tried, true, and fast construction method which we had employed with great success at Log End Cave: the surface bonding of dry-stacked concrete blocks.

The lesson learned from this discouraging story is: With the denser varieties of wood, much less drying time is recommended than with light woods, such as eastern white cedar. My advice now with hardwoods is to strip, split, and stack the wood in single ranks at its final log-end length for just 6 to 10 weeks prior to construction. If in doubt, build a small—say 3' × 5'—test panel. After a few days' curing, subject one side of the panel to a good soaking. If the wood is too dry, the wet side will swell and cause stress-cracking in the mortar joint.

If the log-ends shrink after laying them up, this is mainly an inconvenience, and not a structural problem. Give the wall a year or two to achieve final moisture content and then paint the mortar joints with a thick coat of *pasty* masonry sealer, such as Thoroseal®. The masonry sealer will reach into and fill the shrinkage cracks between wood and mortar. Perma-Chink is also excellent.

BE AWARE

The hardwood expansion problem seems to be rare. The two or three cases that I know that became serious all involved very dry dense wood, and very wet conditions. The infamous crack seems to like to develop just below the first course of wood, usually the result of water collecting on a foundation slab. To review, I feel strongly that the problem can be avoided if one or more of the fol-

12-2. We tested our parging and waterproofing techniques prior to dismantling the wall of dry hardwood.

lowing cautions are exercised:

(1) Do not use very dry hardwoods or dry dense softwoods. If you must use dense log-ends, dry them just 6 to 10 weeks. Better to deal with the relatively minor problem of shrinkage later on.

(2) Keep water from collecting against the first course. Some options here would be: (A) use a starter course of 4" solid blocks, (B) pour footings only (leaving the floor pour until after the roof is on or using a wooden floor system), or (C) coat the first course of wood and mortar with a good clear liquid masonry sealer such as Thoro Corporation's Thoroglaze®.

(3) Use a nonexpansive wood for the first course. Kim Hildreth, you may recall, used northern white cedar on his first course, then switched to the red pine for the rest of the building. He experienced no problem.

(4) Build under cover, if at all possible.

WATERPROOFING CORDWOOD MASONRY BELOW GRADE

Prior to tearing the wall down for the last time, we tested our method of waterproofing cordwood masonry for below-grade applications and found it to be successful. First, we parged the wall with a scratch coat of 3 parts sand and 1 part masonry cement. This coat fills the greatest of the remaining depressions and readies the wall for a good smooth finish coat of the same mix. Using a corner of the trowel, the first coat is scratched with a 3" diamond-grid design, in order to provide a good receiving surface for the finish coat. The W.R. Grace Bituthene® 3000 waterproofing membrane adhered beautifully to the finish coat. A year later, we used this waterproofing method with success at the round cordwood sauna, which is partly earth-sheltered.

13

The Curved-Block Wall

On July 22, we started afresh on the northern hemisphere wall with concrete blocks, although the southern part of the building was almost up to plate height using cedar log ends. The masonry stove construction also commenced about this time, but I will leave the telling of that story to Chapter Fifteen.

We chose 8" concrete corner blocks for our walls. The actual dimensions of these blocks are 7⅝" × 7⅝" × 15⅝", to allow for mortar joints in traditional block wall construction. Except for the first course, however, we would not be using mortar, preferring the surface-bonded block technique for its ease of construction and greater strength against lateral loads. The blocks would be laid widthwise in the wall like cordwood; so we specified *corner blocks only* so that we would not have to worry about the half-cavities found on the ends of regular blocks. My initial estimate showed that we would need at least 1700 blocks for the entire job. We ordered 1100 for the lower storey, as well as a hundred 4" solid blocks and fifty 2" solid coping blocks. The 4" and 2" blocks would give us some latitude in filling spaces at the ends of courses and around the earth tube inlets, described below.

The first course of blocks is laid in a mortar bed, not to "glue" the blocks to the footing, but to establish a perfectly level first course for the dry-stacking of succeeding courses. A mortar of 3 parts sand and 1 part masonry cement is good. No mortar is placed between blocks even on the first course. (*See* 13-1.)

Succeeding courses are dry-stacked, with each block covering the space between blocks on the previous course. On the outside of a 38' 8"-diameter round building, there will be a gap between blocks averaging 0.55", or between ½" and ⅝". This is due to the greater circumference of the outside diameter of the wall. The tops and bottoms of blocks should be cleaned of burrs by rubbing blocks against one another, or by rubbing them with a piece of broken block as shown in 13-2. During stacking, metal shims are often used to keep the wall level and to stop blocks from rocking on two corners. (See 13-3.)

As we stacked the blocks, we filled the cores with sand for additional thermal mass. Filling the cores with concrete would be difficult, time-consuming, and unnecessary for strength. Filling the cores with vermiculite or other loose-fill insulation is not recommended, as this will decrease the value of the thermal mass in the blocks. Surface-bonded walls derive their strength from the application of a tensile

13-1. The first course of blocks is set in mortar.

13-2. Rohan Roy, then 5, cleans the burrs off the concrete blocks, but he advises the use of rubber gloves.

13-3. Dennis checks the wall for plumb and level. Minor corrections are made with metal shims.

membrane to both sides of the dry-stacked blocks. This membrane is composed largely of Portland cement and millions of tiny (½" long) glass fibres. There are other trace ingredients added, such as calcium stearate for waterproofing. The tensile strength (resistance against lateral loads, such as backfill pressure) has been found to be six times greater with surface bonding than with conventionally mortared walls.[8] I use Conproco Foundation Coat surface-bonding cement, which has Type AR (alkali resistant) glass fibres (the only kind acceptable to Massachusetts building codes, incidentally).

Students at Earthwood Building School, still reluctant to let go of traditional building methods in favor of this magical surface-bonding technique, often ask, "Wouldn't the strongest block wall be mortared up first and then surface-bonded?" The answer is, "No." Figure 13-4 shows the join between adjacent blocks. The blocks on the left are mortared together using the standard ⅜" mortar joint. Very few of the ½" glass fibres in the vicinity of the mortar joint actually bond block to block, and those that do, do so only barely. At best, the glass fibres bond each block to the mortar joint, which is the weak link in a block wall. Cracks in a failed block wall invariably follow mortar joints. On the right, where the blocks are stacked correctly without mortar, virtually every glass fibre in the vicinity of the join bonds block to block. This is much stronger.

Actually, the inner surface of the Earthwood block wall is already phenomenally strong against

13-4. The join between blocks.

lateral pressure because of the arch effect already discussed. We used the surface-bonding cement to give smooth walls and a good surface for the application of the waterproofing membrane, as much as we did to gain strength. The wall cannot fall in because the inner surface of the wall is on compression. The wall will not fall out because of the backfill pressure. The surface-bonding cement is overkill, as far as structure is concerned.

On rectilinear construction, however, the story is entirely different. The inner surface of the wall is in tension and the outer surface, therefore, in compression. The surface-bonding cement is a structural necessity.

Prior to surface bonding, we mixed up a grout of the good old 3 parts sand and 1 part masonry cement all-purpose mud, and filled the gaps between blocks on the outside, as well as the troublesome little keyways cast into one end of each block at the block plant. These keyways are designed as an aid in installing doors and windows in normal block construction, but they are a royal pain when using the blocks as we did.

In brief, the surface-bonding technique is:

13-5. Surface-bonding cement is applied with a flat trowel.

(1) Wet the walls to prevent overly rapid drying of the surface-bonding cement.

(2) Mix the bag of cement with water according to the manufacturer's instructions.

(3) Apply the wet cement to the damp wall using a flat trowel, as shown in 13-5. The inner surface of an earth-sheltered wall should have an average thickness of ⅛", which can be spot-checked with the corner of your trowel until you get the knack and flow of things. On the outer surface, an honest ¹⁄₁₆" is good enough for earth-sheltered construction, as the backfilled side of the wall is not in tension, anyway. Above grade, a ⅛" application is necessary on both sides of the wall.

(4) After application, mist the surface-bonding cement frequently, at least twice within the first 24 hours. This will greatly reduce the chance of shrinkage cracking.

All the surface-bonding manufacturers will be glad to supply additional information on their prod-ucts, explaining such things as hot- and cold-weather applications, and giving tips on mixing and applying the cement.

EARTH TUBE INLETS

On the second course of blocks, we left space at predetermined locations for the installation of our *earth tube* (also called *earth pipe* or *cool tube*) system, a series of 4" inlet pipes which allow outside air to come into the house... after it has been moderated by the earth temperature. (*See* 13-6.)

TERMITES

Earth tubes were invented by African termites several million years ago. They also invented the arch, incidentally, and may be counted amongst the planet's first gardeners, but these are other stories. Anyway, it is necessary that the queen's chamber be kept to within 2° Fahrenheit of 70°, in order that she can continue to produce tens of thousands of baby termites every day.

Temperatures in the Zimbabwe savannah range from 100° during the day to 40° at night, but the

13-6. Earth tube inlets are 4"ABS plastic pipes.

queen's chamber, near the center of the mound, stays at 70°. The trick? A thermal chimney and a system of earth tubes. It works like this: The opening at the top of the mound—sometimes 20' high—is orientated to the south or southwest. This opening—the thermal chimney—is connected by a main shaft to deep within the mound, where it divides into several smaller tubes which make their way in different directions to various earth-tube openings quite a distance from the mound. The queen's chamber is supported by columns and arches at the center of the main tube. In the sweltering afternoon, the sun heats the air in the thermal chimney at the top of the mound. (Again: *Warm fluids rise.*) The easiest route for return air is through the earth tubes, which moderate the incoming hot air by contact with the cool earth beneath the surface. The queen's chamber is constantly filled with earth-tempered air and thereby maintains a moderate temperature. At night, the system works in reverse. Air warmed by the mound's mass rises, drawing cool air into the earth tubes, where it is warmed to 70°. The Earthwood design was intended to work like a giant termite mound, but termites are better engineers.

THE THERMAL CHIMNEY

Bluntly, our thermal chimney, seen in 15-13, didn't work. The construction of the chimney was two-by-twelve vertical planks alternating with two-by-six-teen planks, giving internal dimensions of 12" square by 4' high. The southeast and southwest sides were glazed with single panes of ¼" plate glass, the taller pane on the southwest for greater solar gain in the hot afternoon. The interior surfaces of the chimney were painted black for solar absorption. On a hot sunny day of 88° F, the interior of the solar chimney would not rise above 106°, only an 18° differential. The rising draught of hot air was insufficient to draw outside air in through the earth tubes, even though all the doors and windows were closed. Embarrassingly, the earth tubes were actually emptying the cool air out of the lower storey. This phenomenon makes perfect sense when no thermal chimney is in use, as there is nothing to stop the naturally cooler—and heavier—air at the bottom of the building from spilling out of the earth tubes.

The idea, of course, was that a sufficient upward

draught would be created by the thermal chimney to reverse this flow, but, alas, we could not derive a temperature differential of more than 20° F, despite several experiments with changing the area of the top vent opening, increasing thermal mass inside of the chimney by hanging black steel absorption plates, and increasing solar gain by changing the chimney cap to glass.

Still unsure whether the problem was with the thermal chimney or the earth tubes, we conducted an experiment with an electric fan to draw air out of the top of the chimney. Despite 87° outside temperatures, cooled air was being drawn into the lower storey through the earth-tube inlets at temperatures ranging from 61° to 66° F, depending upon the location of the inlets. Those on the north side of the house performed the best, as would be expected. The earth tubes were capable of cooling air by 21° to 27° F, a dramatic result.

What we learned by this was that our thermal chimney was woefully inadequate. But we also learned that Earthwood stays quite cool on its own, even during a stifling summer. The upstairs temperature would rise to 79° on the very hot days, while the lower storey remained at 66° through July and about 69° in late August. All we need to do at Earthwood is to destratify the air, both in summer and winter. The easiest way to do this is with a Casablanca-type ceiling fan positioned over the circular stairway. An alternative would be to install a fan in an upstairs window on a permanent basis.

Finally, there is no doubt in my mind that a solar chimney can be made to work, but it is obvious that it would have to be very much more substantial than the one we tried. For our part, we feel that Earthwood's performance at this latitude does not necessitate the use of a thermal chimney, and the lines and symmetry of the house are greatly improved by its elimination. The earth tubes, though, have a value of their own. They help to ventilate the otherwise stagnant corners of the lower level and they work very well in preheating the incoming winter air used in stovewood combustion.

SIMULTANEOUS EVENTS

In a book, it is convenient to talk about one mode of construction at a time, but, in reality, several components of the structure progress together—along with the various ancillary systems. This has to be so, as surely as the first storey must be completed before the second storey can be built. At this point in time at Earthwood, we were actually attending to cordwood construction, block walls, post-and-beam frame-work, the masonry stove construction, some water-proofing (as the walls reached first-storey height),

and footing drains. Our footing drains are a part of the earth-tube system, for reasons explained in Chapter 16 about drainage. Suffice it to say at this point that 2'-1ong, 4"-diameter, heavy ABS plastic pipes were laid up in the second course of blocks, with 8" protruding on the exterior for later connec-tion to the footing drains. Space around the pipes was filled with cement.

14

Post-and-Beam & Plank-and-Beam Framing

We will begin this part by clarifying terms. The easiest way to do this is to change the *beam* of *post-and-beam* to *girder*. This is the name for the heavy timber which runs from one post to another. The girders support the *beams* of the *plank-and-beam* floor or roof systems. These beams are more accurately called *floor joists* if a floor is supported, and *rafters* if a roof is supported. So now we have *post-and-girder* and *plank-and-rafter* (or *joist*) systems. Refer again to 8-1 and 8-7.

ON POSTS AND PLANKS

We will simplify the discussion further by dispensing with the strongest components of these systems first. Posts (also called *columns*) are phenomenally strong. For example, an 8'-high, eight-by-eight post of a wood with relatively low compression strength of 1150 to 1400 inch-lbs per square inch will support 63,000 lbs.[9] Our earth roof—fully saturated—weighs "only" 207,750 lbs, maximum. (1385 sq ft × 150 lbs pre sq ft = 207,750 lbs.) Sixty-eight percent of that load is carried by the walls and center stone mass, and no single post carries more than 6 percent of the total load, or 12,465 lbs. An eight-by-eight post is five times stronger than it needs to be. Even a six-by-six post would be overbuilt by a factor of 2.6, although I would not recommend a column smaller than six-by-six (or a 6"-diameter round post), because of something called *slenderness ratio*, which describes the tendency of a tall, narrow post to snap against lateral pressure.

The other strong component is the planking, and I recommend dry two-by-six tongue-and-groove spruce, pine, or their equivalent. This planking can easily cover rafters set 4' on center (4' 0" o.c.) and still support loads greater than those on a reasonable earth roof. Before exceeding the carrying capacity of the planking, there will be grave difficulties with the rafters.

DESIGN CONSIDERATIONS

It is outside of the scope of this book to enter into the calculations necessary to design one's own plank-and-beam system for supporting an earth roof. Be aware of these five considerations which have to be integrated:

(1) *Load:* An 8" earth roof, fully saturated, and with a 50-lb snow load will weigh approximately 150 lbs per sq ft. This is three times the load expected with ordinary frame construction.

(2) *Kind of wood:* Each species and grade of wood has a different unit stress rating for shear and bending. Know your wood. There are pages and pages of grading tables in engineering manuals.

(3) *Frequency of rafters:* What is the on-center (o.c.) spacing? Rafters 2' apart (center to center) are called 24" o.c.

(4) *Dimensions of rafters:* Four-by-eight? Five-by-ten? 8"-diameter logs?

(5) *Span:* What clear span do you wish to support? The strength of a rafter or girder is inversely proportional to the square of the span. This is very

important. It means that a rafter has to be four times as strong to support a 16' span as to support an 8' span—other things being equal—not merely twice as strong, as might be expected.

You must know four of the five variables listed above to calculate for the fifth. If you know *load, kind of wood, rafter frequency,* and *span,* for example, you can calculate the cross-sectional dimensions of the rafter. Or, given the kind and grade of wood, you can calculate what load a particular rafter system can support.

BENDING AND SHEAR

There are two different kinds of failure in girders and rafters with which we should be concerned: bending failure and shear failure. Bending failure occurs when a rafter breaks near the center under a vertical load. Rafters are usually much taller than they are wide in order to provide greater bending strength per board foot of lumber. This is because of the characteristic of cross-sectional rafter shape called *section modulus* (S) which is used with bending formulas. Now, section modulus is a pretty scary-sounding term, but don't worry. It does not take a degree in higher mathematics to see the importance of shape in the section modulus formula for beams of rectilinear cross section: $S = bd^2/6$. The letter b corresponds to the *breadth* of the beam, *d* refers to its *depth*. Note that the value of the depth of the beam is squared, while the breadth is not. The 6" × 12" beam shown at the left in 14-1 has a section modulus of 144 inches cubed when installed in the regular fashion (S = 6" × (12")²/6 = 144 inches cubed,) but only 72 cubic inches when laid up incorrectly, as shown on the right. (S = 12" × (6")²/6 = 72 inches cubed.) This makes sense, to us, because we've seen it so often, but it's nice to know that common sense is backed by fact.

With shear strength, however, we are concerned with the cross-sectional area (A) of the beam, not the section modulus. The formula A = bd places equal value on breadth and depth. The beam in 14-1 has a cross-sectional area of 72 square inches, regardless of which way it is laid. Shear strength is the tendency of a beam to "shear off" right near one of its supports, as seen in 8-6. The actual numbers of wood fibres which must shear is the critical thing, so only

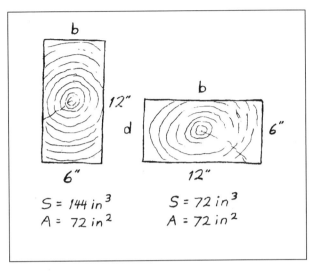

14-1. Comparing cross-sectional characteristics of two beams installed in different ways.

cross-sectional area is important. Round rafters, often found in traditional buildings in the southwest United States, make efficient use of logs with regard to shear strength. With ordinary light-frame construction, bending strength seems to come into play before shear strength, but on heavy-timber construction required for very heavy loads, the opposite is quite often the case. Look at 14-2.

It is true that there will be less "sag" (deflection) in the lower example. But deflection does not matter much on an exposed plank-and-rafter ceiling, as there is no plaster or spackling to crack and fall down. And, in point of fact, the example at the top of the drawing is actually stronger with regard to shear. With heavy-timber construction, it is most often the case that when all the pencil pushing is done, shear strength is more of a concern than bending strength. The mathematics of this are quite involved, but any structural engineer will verify that this is so. The thing to remember is that shear strength can be increased by using beams of greater cross-sectional area. This will be of particular interest to those readers who have considered using their own standing timber for their rafters. The logs can be stripped of bark and flattened on one side with an adze for easy joining and nailing of planking. No sawmill is required.

The upshot of this discussion is that your post-and-girder and plank-and-rafter systems should be

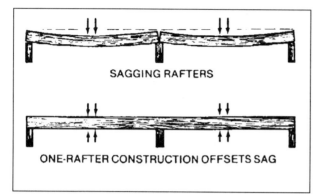

SAGGING RAFTERS

ONE-RAFTER CONSTRUCTION OFFSETS SAG

14-2. Deflection is decreased with a double span, and bending strength is increased... but shear strength is decreased about 25 percent.

carefully checked by a competent structural engineer prior to building the house. His trained eye may spot some other glaring design fault. I have tried to impart a sense of some of the trade-offs involved in this kind of design work so that the reader has more options available to correct any stress-load deficiencies in the design or to better balance the use of materials when a design is grossly overbuilt. Underbuilding is disastrous. Overbuilding is unnecessarily expensive. Therefore, retaining an engineer is both prudent *and* economical.

THE EARTHWOOD FRAMEWORK

The easiest, and, in my opinion, best way to lay out joists and rafters in a round house is by the radial rafter system, where one end of the beam is supported in the center by a column and the other end supported by the exterior wall. At Earthwood, however, a 16' clear span for either joists or rafters is much too great. Remember the relationship of span length to strength, discussed above. A post-and-girder system supporting the rafters halfway is required. Either a hexagonal or octagonal configuration can be used here. We chose the octagon because of the shorter girder spans, although we designed out one of the posts as already mentioned, which necessitated the use of one extremely powerful girder: ten-by-twelve clear oak.

ON OLD BARN BEAMS

We used old barn timbers for our posts and girders, except for the new oak beam. We like the character of the old hand-hewn beams. Although posts are extremely strong—and we used nothing smaller than seven-by-seven—girders are another matter. If new timbers were cut from Douglas fir or an equivalent wood, eight-by-eight girders would be adequate. But as we were using old timbers, even though we checked them with extreme care for deterioration, I decided to use nothing less than ten-by-ten for the three clear spans upstairs supporting the heavy roof rafters. Downstairs girders, which only support the relatively light floor joists and planking, were overbuilt, using eight-by-eight barn timbers. An eight-by-eight girder was also used where an internal wall helped to support it from below. These internal walls upstairs are always superimposed exactly above corresponding walls downstairs, except where the heavy oak beam negates the need for additional support.

Which old barn timbers to use is a value judgment on my part and I am very fastidious in their selection. We find plenty of uses around the homestead for rejects. Our strategy, which we've used at all three houses that Jaki and I have built, is to carefully examine and catalog each individual beam, noting its dimensions and quality. The framework is designed on paper using this list as a guide. If we're a beam or two short, we seek out replacements; we do not try to make a questionable timber serve the purpose.

If you're building in a coded area, the use of recycled timbers for structure may not be approved by the building inspector. New rough-cut posts and beams from a local sawmill are very strong, economical, and can be quite attractive, especially if they are bleached after a year or two in the building.

ERECTING THE FRAME

On a concrete floor, it is advisable to put a vapor barrier (also called a *damp-proof course*) between the posts and the concrete. Although any waterproof barrier will probably do for this purpose, I like to use squares cut from heavy roll roofing or asphalt shingles. A 24"-square piece of scrap will yield nine 8" × 8" squares. I

14-3. Girders made from old barn beams support the four-by-eight floor joists. The girders are fastened to each other with truss plates. Dennis checks that rafters are level with each other.

like the heavy roofing material because it helps average out irregularities in the saw cut upon which the post is standing. The post is much steadier.

Dennis and I cut and set the seven posts and six of the girders on June 10 and 11. The seventh girder was the big oak, still at the sawmill, so we installed a temporary two-by-six as a substitute to give our frame some rigidity. People are sometimes surprised to hear that we don't fasten the posts to the floor in any way. Believe me, with six tons on each post, they aren't going anywhere. If you are squeamish about this, you can go to the trouble of drilling a hole in the concrete floor and another in the bottom of the post, and joining them with a metal pin. I don't like the idea of setting anchor bolts in the floor during the pour, as they get in the way of trowelling, people will hurt themselves tripping on them, and positioning pins make it difficult to make minor adjustments in plumb.

The girders are fastened to the tops of the posts with toed nails, and fastened to each other with truss plates, as shown in 14-3. We carefully planned the post lengths and girder thicknesses so that the top of the girder ring would be 7' 3½". This corresponds exactly to an exterior wall of 11 courses of 8" (actually 7⅝") blocks and a single course of 4" (3⅝") solid blocks. The floor joists would be level.

The floor joists are four-by-eight hemlock. There are 16 primary joists—each at least 18' long—which are supported by the center stone mass at one end (see Chapter 15), the octagonal post-and-girder system near the midspan, and the walls at the other end. In point of fact, 11 of the 16 primary joists are about 21' long. The extra 3' cantilevers over the wall to support the walk-around deck. In addition, there are 16 secondary joists which do not extend to the center. Planking spans become so short within the octagon that every second joist can be eliminated. You may wish to refer again to 8-1.

The joists rest on the block wall with only a damp-proof course of plastic between joist and block. Remember that the last course of blocks before the floor joists is composed of 4" solid blocks. On the southern half of the house, floor joists are supported by the cordwood walls, which have had wooden plates set in the masonry expressly for the purpose.

PLATES

A system of 2" × 6" plates is set in the cordwood wall at such a height that the joists will be level; in our case, the tops of the plates are 7' 3½" off the floor. The same system is employed later to support th roof rafters. For the roof rafters, there is a full complement of 32 pairs of plates, corresponding to the 32 rafters. There are fewer pairs of plates supporting floor joists because of the block wall on the north.

Exterior plates are cut at 3' 7" and internal plates at 3' 5" to accommodate the external and internal circumferences, with a little play between sets for ease of placement. Where there are window or door lintels, the plates are nailed right to the heavy lintels, which themselves must be set at the right height. A thin layer of fibreglass insulation is compressed between the plates and the lintels during this nailing, as shown in 14-4 and 14-5. When fastening plates directly to cordwood masonry, we set 8 or 10 roofing nails in the underside of each plate, leaving them sticking out about ¼" to help grab the mortar. Otherwise, there is a very much greater chance of the plates being knocked loose during the joist installation.

Some cordwood builders lay up floor joists and rafters right into the wall without benefit of plates. I have never been too enthusiastic about this, especially with heavy rafter loads. The plates help distribute the load, preventing a slicing effect into the cordwood masonry.

Joists and rafters can be levelled to each other with shims made from wooden shingles. I have found this to be preferable to *bird's-mouthing* because it is much easier and does not cut down on the cross-sectional area of the beam, nor therefore. the shear strength. (*See* 14-6.)

PLANKING

The deck strength at Earthwood is predicated upon using two-by-six tongue-and-groove planking, pretty much standard in earth-sheltered housing having a wooden substrate. If the boards are V-jointed on one side, always keep this side to the bottom.

Laying planking on a radial rafter system is not really any more difficult than with parallel rafters, although it does take more time. On the positive side,

14-4 (left) and 14-5 (right). To eliminate drafts, Jaki installs a strip of fibreglass between the lintels and the plates.

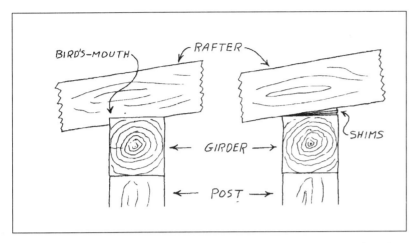

14-6. Left: Cross-sectional area is lost—along with some shear strength—when a rafter is bird's-mouthed to rest solidly on a girder (or wall plate). Right: Shimming does not decrease the shear strength of the rafter.

14-7. Jonathan Cross nails the two-by-six tongue-and-groove spruce planking to the radial rafter system. Even short pieces are useful.

there is very little waste, because good use is made of short scraps near the center, as shown in 14-7.. Start at the outside and work in towards the center, one facet of the floor—or roof—at a time. One end of the board needs to be fitted to the edge of the previous facet—use a bevel square to get this angle right—but the other end can be left long. When the facet is covered, snap a chalkline over the center of the joist and cut all the long boards off cleanly with a circular saw. Just set the depth of the saw about ¹⁄₁₆" deeper than the thickness of the planking. It is important to keep your nails well clear of the center of the rafter, or you will hit them with the saw. We found that 12-penny galvanized cup-headed nails hold well and can be countersunk easily to be out of the way of floor sanding.

To save on waste, take time to select the piece nearest the length you need, erring on the long side. After a while, you will learn how to make the most of your boards. If you start with all 12' boards, for example, you will soon discover that the cutoffs can be used in reverse order further down the facet. With care, there will be no waste over a foot long, except for damaged boards.

THE ROOF

There are only three differences between the roof system and the floor system: (l) Rafters are five-by-ten, not four-by-eight, except that seven at six-by-ten are used for the longer spans in the bedrooms. (2) Much more decking is used because of the roof overhang. (3) The roof is pitched. We used a "one-in-fourteen" (1:14) pitch at Earthwood, but any pitch between 1:12 and 1:18 is good for an earth roof. Water will not pond on the membrane and the earth will not slump towards the edge. (A 1:12 pitch means that a rafter rises 1" for every 12" of its length.)

15

The Masonry Stove

Eastern Europeans say that Americans don't know how to burn wood, and I'm inclined to agree with them. Most Americans know that fireplaces are hopelessly inefficient, and will actually empty the heat out of a house in the wee hours when the fire goes out, thanks to the draught established in the chimney flue. (This loss can be greatly reduced by the installation of airtight fireplace doors.)

Modern wood stoves are a tremendous improvement, of course, but there are still problems. In an effort to keep the fire alive overnight, the airtight dampers are closed and combustion slows down greatly. Unfortunately, the firebox temperature drops below that required for complete combustion, causing unburned volatile gases to start their way up the chimney where—even more unfortunately—they condense in the form of creosote. Dirty and dangerous. Enter the masonry stove, also known as a "Russian fireplace" (15-1).

It works like this: The stove is fired up once a day—the right time will be determined by experimentation—and an armload of wood is allowed to burn extra hot, that is, with plenty of air coming through the damper control. Perhaps a second charge of wood is loaded after an hour or two, perhaps not—this depends on the time of year. While the wood is being burned at high temperature, all the volatiles are combusted and there is no creosote buildup.

When all the yellow flames are gone, and only clear flames are coming off the hot coals, the blast gate is closed and the fire is allowed to transfer its heat to the stone mass in its own good time. The extremely long flues, crisscrossing throughout the

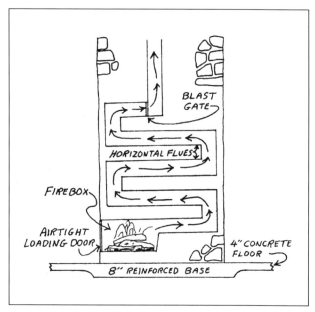

15-1. The heat normally lost out of a straight chimney is absorbed by the mass in a masonry stove, due to the increased surface area of the flue. When only hot red coals are left, the blast gate is closed.

masonry, provide lots of surface area for the transfer of the exhaust heat to the thermal mass. If properly designed and built, the exhaust into the atmosphere is cooled enough so that you can hold your hand comfortably over the chimney top—*and* virtually no pollutants are put into the atmosphere.

The tremendous mass of the stove stores the heat, giving it off slowly to the house interior.

Twenty-four hours later, the mass is recharged with a new fire. With masonry stoves, less fuel is used, the fire burns cleaner (little ash is left) and the temperature is steady. The only drawback is that more kindling wood must be stocked. However, this is offset by the advantage of having to fire the stove and clean the firebox much less frequently.

In a true integration of systems, the masonry stove at Earthwood is part of the same 21-ton stone pillar which supports the inner end of the floor joists and roof rafters. This should only be done with an extremely massive masonry stove, such as ours, as temperatures may be reached in a smaller stove which might begin to "crisp" the floor joists. Some building inspectors may frown on the Earthwood plan because of regulations concerning "clearance from a chimney to combustible materials" or some such. The reality is that the stone shelf upon which the floor joists rest never exceeds a temperature of about 100° F. The shelf where the roof rafters rest does not exceed room temperature.

Our masonry stove is cylindrical, with a diameter of 5' on the first floor, reducing to 4' through the second storey. This reduction in diameter forms a 6"-wide shelf around the stove for resting one end of the floor joists. Likewise, the diameter is reduced by a foot again as the masonry mass passes through the roof. Another 6" shelf is provided for the rafters.

CONSTRUCTION OF THE EARTHWOOD STOVE, STEP BY STEP

The photographs that follow are not intended to be a definitive and complete work on masonry stoves. A book would be needed for this. Building a stone masonry stove is best left to an experienced stone mason, although a brick-faced stove would probably be within the capabilities of a talented owner-builder. The intent of this section is to show some of the special features of the Earthwood stove and its integration with the house structure.

After drawing the shape of the stove on the footing with a red crayon, mason Steve Engelhart begins the stonework. (See 15-2.) The mortar mix is 12 sand, 4 masonry cement, 1 lime—mixed quite stiffly for stone masonry. The firebrick floor is about 9" off the floor for ease of loading.

Steve builds the firebox of firebricks, using refractory cement as the bonding material. (See 15-3.) The space between the stone facing and the firebox is filled with "rubble" stones and mortar. Don't waste your good building stones here.

The firebox is approximately 26" high, 33" deep, and 15" wide, interior measurements. The arch form resting on top of the brick sidewalls is made of wooden cross-members and heavy aluminum flashing. (See 15-4.) It is left in the construction, where it will burn out later. The exhaust from the firebox is to the rear.

This firebox is a little too big. In order to shape the red-hot coals into a more compact mass—which lasts longer—I have lined the lower portion of the firebox walls with additional firebricks standing on end. A smaller firebox of 24" high, 30" deep, and 12" wide would have been better. See also "IF I WERE DOING IT AGAIN..." at the end of this chapter.

The firebrick arch is built, with the bricks laid up on their narrow sides. The arch is capped with mortar reinforced with metal lathe or ¼"-mesh hardware cloth. The doors are from an Oval airtight cookstove, but any airtight cast iron stove doors will do. (See 15-5.)

The first 8" × 12" ceramic flue extends out the back of the firebox, laterally, with a rectangular opening already cut into its side with a masonry saw. Note the flue tile in the background, already cut in a similar fashion. A short flue section carries the chimney upwards to the next course. Two 24"-long flue tiles run horizontally over the firebox, supported by old bricks. (See 15-6.)

The second long run of the flue is checked for level. (See 15-7.) At one end of each run, we install a circular 8" ceramic thimble to provide a clean-out. The clean-outs can be blocked off with some fibreglass insulation and an 8" stovepipe termination cap. It is important to be able to monitor the inside of the stove, in case there is ever any kind of blockage from some unexpected source.

A strong cap of Portland cement (3 sand, 1 Portland) is built on top of the lower-storey mass so that the top of the cap corresponds to the underside of the floor-joist system. (See 15-8.) The stonework now reduces to 4' in diameter. Note the clean-outs, one covered with a termination cap.

The blast gate, looking a bit like a mail slot, is located in a vertical run of flue in the upstairs portion of the house. (See Illustrations 15-9 and 15-10.) A heavy (¼") metal plate slides back and forth in the

STEP-BY-STEP CONSTRUCTION OF THE EARTHWOOD STOVE

15-2. Drawing the shape of the stove.

15-3. Filled with "rubble" stones and mortar.

15-4. Line the lower portion of the firebox walls.

15-5.The firebrick arch.

15-6. Flue extends out the back of the firebox.

15-7. Flue is checked for level.

15-8. Cap of Portland cement.

15-9. The blast gate.

15-10. Horizontal flue clean-out.

15-11. Mass provides a shelf for the roof rafters.

15-12. Galvanized flashing cone.

15-13. Chimney continues for another 32".

slot, accessible by a hook built into the plate. In the closed position, very little heat is lost up the chimney. Upstairs, there are only two horizontal flues, each accessible by a clean-out.

The mass reduces to 3' in diameter, to provide a shelf for the roof rafters. (*See* 15-11.) The mass is kept round and fairly smooth by forming it with strong aluminum flashing, which will take a true cylindrical shape when filled with masonry and mortar. (The flashing is held together with several windings of strong duct tape.) This section of the chimney is surface-bonded to make it even smoother.

The importance of the round smooth cylinder is now apparent. A galvanized flashing cone, specially fabricated at a local sheet-metal shop, flashes the chimney to the deck. It is nailed to the deck every 2" around the perimeter, using a rubber washer beneath each head. Bituthene® caulking oozes out the edge. The top part of the flashing cone is sealed with silicone caulking. (*See* 15-12.)

A 3'-diameter stone chimney continues on for another 32", terminating in a strong 4"-thick Portland cement cap. (*See* 15-13.)

PERFORMANCE

The stove has been a great success for ten years. In all that time, absolutely no creosote has formed in any part of the flues or chimney. There is only a small quantity (¼") of "fly ash," which we have removed once, just prior to our tenth heating season. The house stays a steady temperature, summer and winter. I think that the stone mass, a giant capacitor, helps to keep the house cool in the summer.

We continue to use about 3¼" full cords of wood at Earthwood each year for all purposes. About half of this goes through the masonry stove. Using the masonry stove more (and cutting the use of the upstairs cookstove down correspondingly) reduces temperature stratification between the upstairs and downstairs. With wood use split about 50-50 between the upstairs and downstairs stoves, the upstairs is typically eight degrees (8° F) warmer than downstairs. This works out fine with our use of the space.

A good clearing house for plans and information on masonry stoves, and specialized components like blast gates and heat-proof glass doors, is: Maine Wood Heat Co., Rt. 1, Box 38, Norridgewock, ME 04957.   A detailed slide show on the Earthwood stove is available from Earthwood Building School.

IF I WERE DOING IT AGAIN...

I would make two changes in the masonry stove design. (1) I would design the firebox of a shape conducive to hanging a pair of heatproof glass doors. The stove could actually double as a pleasant fireplace, without compromising efficiency. (2) I would follow the advice of masonry-stove advocate Albie Barden, who says:

> Fireboxes should be free to expand and contract within the common brick heat exchange shell and should therefore be completely surrounded with an expansion joint, best kept open with a ¼" to ½"" blanket of mineral wool. [10]

In fact, I would extend the use of the mineral wool to include the first two horizontal flues. We had some expansion cracking in our masonry near the firebox. Steve and I expected this—and it certainly does not constitute a hazard in our case—but the mineral-wool expansion joint would no doubt lessen, if not eliminate, this minor cracking.

16

Waterproofing and Drainage

One of the first questions asked when the subject of earth-sheltered housing comes up is, "What do you put on the roof (or walls) to keep the water out?" People want to know which membrane to use, as if this is the magic stuff which keeps an underground house dry. In reality, what keeps an earth-shelter dry is our old friend, gravity.

Waterproofing and drainage go hand in hand. A good shingled roof does not leak because gravity draws rainwater from one shingle to the next, finally depositing the wet stuff into a gutter. Drainage is obviously the key here, not watertightness. A flat shingled roof would fail miserably.

The same is true below grade. Any waterproofing membrane, no matter how exotic, is subject to potential failure. We should not try to dam the water out of our houses. Rather, let us take the more gentle and effective approach of carrying it away from the building, thus reducing the pressure on our waterproofing to a minimum.

In *Underground Houses*, I described in detail an effective and cheap (albeit dirty and time-consuming) method of waterproofing which involved bedding layers of 6-mil black polyethylene into a trowelled-on layer of black plastic roofing cement. We had no membrane leaks at Log End Cave by using this method, although I credit good drainage techniques for this more than the quality of the membrane. Moreover, the application was a long and tedious job which I would not wish upon people I want to keep as friends. I would recommend it only to someone who is trying to build a home on a materials budget of under $8 per sq ft.

The membrane which I have come to love and

respect over the past ten years provides a superior protection of a similar kind to the black poly method, but it is infinitely easier and quicker to apply—fully twenty times quicker. It is the Bituthene® waterproofing membrane, a tough, pliable, self-adhesive material composed of high-strength polyethylene (two layers, cross-laminated like plywood), factory-coated on one side with a uniform layer of about $\frac{1}{16}$" (0.060") of rubberized asphalt. This bituminous bedding is very sticky, but the 3' × 60' roll comes protected by a layer of non-stick backing paper, which is removed while the membrane is applied.

At Earthwood, we used the Bituthene® 3000 membrane for both walls and roof. The applications are a little different, but I will cover both now.

WATERPROOFING THE WALLS

After the surface-bonding cement has cured, the wall is primed with the Bituthene® P3000 primer. This product comes in a 5-gallon pail and is applied with a lamb's-wool roller. The primer is allowed to dry one hour or until it is no longer tacky. It goes on very fast.

Before applying the main sheets of membrane, we attend to the critical detail where the wall meets the footing. A strip of Bituthene® about 7" wide and 36" long is cut from the end of a roll. The backing paper is removed and the strip is carefully fitted into the crease where the wall and footing meet. About 3" of the strip is now stuck fast to the footing, and the

186

16-1 (left) and 16-2 (right). The important detail where the footing meets the wall is sealed with a strip of Bituthene®.

rest to the wall. The edge stuck to the footing is sealed for additional protection with a bead of a compatible Bituthene® EM3000 mastic. (*See* 16-1 and 16-2.) The bead of mastic is applied with a caulking gun, and then smoothed with a pointing knife. All freshly cut edges of the membrane are sealed in this way, but it is not necessary to apply mastic to the factory edge.

Sheets are applied to the wall vertically, as shown in Illustration 16-3. Cut the sheet to the right length, and have someone hold the top edge from above. Two others work from below, one removing the backing paper, the other pressing the Bituthene® onto the wall. Lap the previously laid strips at the footing detail with a full 3" overlap, and seal the bottom with mastic. Adhesion is best begun at the middle of the sheet's width and spread out right and left with your hands. This will prevent any bubbles from forming behind the membrane. Continue pressing the membrane into place right to the top and discard the backing paper. The second 36"-wide piece laps the first by 2½". A yellow lap line is provided on each

16-3. The Bituthene® waterproofing is applied vertically to walls.

16-4. The lower storey of Earthwood is waterproofed.

edge of the sheet to aid in this. If it becomes apparent that the second piece is wandering off the line, stop pressing the Bituthene®. The sheets cannot be pulled or stretched back to alignment. Cut the sheet short at this point, and begin again with an overlap of 3" at the bottom. Use heavy hand pressure at the lap seam and seal the edge of the lap join with the compatible mastic. Never use an incompatible mastic, such as tar, asphalt, or pitch-based materials, or mastics containing polysulfide polymer.

The beauty of the Bituthene® membrane is the speed and ease of application. Any shape of piece can be cut to fit, but always remember to allow for the 2½" or 3" lap and to seal all cut edges. Illustration 16-4 shows the waterproofing complete on the lower portion of the block wall.

WATERPROOFING THE ROOF

Before waterproofing, the edge of the roof must be flashed. We used a 10"-wide flashing, bending 3" over the edge of the roof to hide the edge of the planking and to act as a drip edge. Seven inches of the flashing rests on the roof. Nail flashing with galvanized roofing nails about 4½" in from the edge of the building. The Bituthene® will cover these nails. We laid our first course of Bituthene® so that 3" of the flashing on the roof surface was left exposed.

The Bituthene® membrane can be applied directly to seasoned planking which will not be subject to great shrinking. If green planking is used, a layer of some inexpensive hardboard material such as ¼"

plywood underlayment should be laid down first. Prime the planking with the Bituthene® P3000 primer for improved adhesion. At Earthwood, the roof is composed of 16 identical facets, something like the blunt end of a diamond. Start at the bottom of the roof and cut a trapezoidal piece of Bituthene® so that it will lap onto the next facet at each end by 2½" This means that the cut edges of the sheets overlap each other by 5" on the roof. Sheets are cut with a knife over a scrap board, as seen in 16-5. Two people must turn the piece over and remove the backing paper. Now, carefully turn the sheet over again so that the sticky side is down and position it correctly with regard to the flashing and/or the edge of the previously laid membrane. Talk to each other. "I'm four inches over!" "I'm only an inch over. Give me an inch and a half!" A third person is very useful in starting the adhesion at the middle, especially with larger sheets. Four-foot-long pieces are easily handled by two people. (*See* 16-5.)

Three people working together will soon get into a rhythm, with one person unrolling and cutting the next piece while the others remove backing paper and position the first. The roof is easily waterproofed in a day by inexperienced workers. After the entire first course has been placed around the edge of the building, proceed with the second course, allowing the 2½" lap as indicated by the yellow line. Factory edges need not be sealed with mastic, but cut edges do, as shown in 16-6.

Punctures, cuts, or other suspicious marks on the membrane are easily taken care of with a patch cut of the same material. The patch should extend 3" in every direction from the damage, pressed hard to the

16-5. Use a scrap board as a backing for cutting Bituthene® with a razor blade knife.

16-6. Apply the lower edge of the Bituthene® first, following the yellow guideline.

surface with the heel of your hand. The edges of the patch are sealed with mastic. I have a habit of making an octagon-shaped patch with my razor-blade knife, so that there is less chance of a corner lifting.

W. R. Grace makes another almost identical product called Bituthene® 3100. The rubberized asphalt is much stickier, and the 3100 membrane

can be used at temperatures between 25° F and 40° F. We used some on the roof at Earthwood, when the October temperatures were getting quite nippy. Because of its stickiness, the Bituthene® 3100 is also recommended for vertical-wall applications. The costs of the 3000 and the 3100 are about the same, which, in March of 1983, worked out to about 60–65¢ per sq ft. With primer and mastic, the total cost came to about 75¢ per sq ft.

16.-7. Nonfactory edges must be sealed with Bituthene® - compatible caulk. Note the aluminum flashing at the edge of the roof.

OTHER WATERPROOFING

There are other options in waterproofing, and many good—and not so good—products on the market. There are some excellent sheet membranes, including neoprene, butyl rubber, and EPDM, a synthetic. These membranes are frequently used in large commercial applications, but they are not easily installed by inexperienced owner-builders—a 40' × 40' sheet of 1/16" material weighs between 600 and 750 lbs, depending on kind—and their installed cost is quite high, about twice the cost of Bituthene®.

A popular waterproofing material is bentonite clay. The clay is refined and manufactured into various products, including panels of corrugated cardboard with fine bentonite in the *flutes* (sluices), trowel consistency mixtures, and spray-on products.

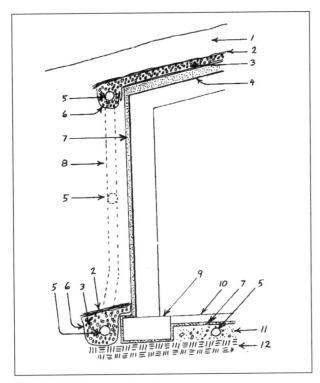

16-8. (Not drawn to scale)

1. Earth.
2. Filtration mat; can be hay or straw.
3. Crushed stone.
4. Waterproofing membrane.
5. 4" perforated drain wrapped in a filtration fabric.
6. 6-mil polyethylene to form underground "gutter." The top gutter should include a fold in the plastic as shown, to allow for earth settling.
7. Rigid-foam insulation.
8. Vertically placed 4" nonperforated drains connect the horizontal perforated drains. T-junctions are available to make these connections.
9. Footing.
10. Concrete-slab floor.
11. Compacted sand, gravel, or crushed stone.
12. Undisturbed subsoil or heavily compacted pad, as previously described.

The panels are well spoken of by builders who have used them, and their cost per sq ft is a little less than Bituthene®. However, the panels must be glued to a surface-bonded wall with a mastic, which adds somewhat to both the cost and the installation time.

The panels are unacceptable for direct application to a wooden deck. American Colloid Company, makers of Volclay® bentonite panels, recommends that a concrete topping be poured over the deck prior to waterproofing. Also, the company does not encourage application of their system by inexperienced owner-builders.

A tube of dry granular bentonite called Hydrobar® is available from American Colloid. This product is an excellent means of arresting water penetration at that critical spot where the wall meets the footings.

All bentonite clay waterproofing works the same way. A layer of bentonite is held fast against the wall or roof in some way. Water approaching the building hits the outer molecular layers of the bentonite, which swell up to fifteen times their original volume, closing off any further water penetration.

The best overall commentary that I have seen on waterproofing, including drainage considerations, is found in the Underground Space Center's book, *Earth Sheltered Residential Design Manual*. See the Bibliography. For ease of application, quality, and cost, I have been very happy with the Bituthene® membrane and do not hesitate in recommending it to owner-builders. Write: W. R. Grace Co., 62 Whittemore Ave., Cambridge, MA 02140.

DRAINAGE

Waterproofing was discussed before drainage because its installation occurs first. But remember, drainage is your main line of defense against water. Give the water an easier place to go than into your house, and it will cooperate with you. The keys to good drainage are the use of backfill with excellent percolation characteristics, and a footing drain (French drain), which carries the water away from the house. At the Log End Cottage basement, we installed footing drains correctly, but backfilled with the poor claylike earth which came out of the hole. No good. The water couldn't get to the footing drains, and hydrostatic and frost pressures caused a movement in the block wall which negated our cement-based waterproofing. (Cementitious waterproofings are not recommended by proponents of earth-sheltered housing because they have no ability to bridge even the tiniest shrinkage or settling cracks.)

A good drain system for an earth-sheltered house is shown in 16-8. With a freestanding roof, such as at Earthwood, a gutter system is advisable, especially in wet maritime climates. This eliminates the need for the topmost drain just below the surface. The intermediate drain is optional, and is only necessary in the case of soils with moderate percolation characteristics. If good percolating backfill, such as coarse sand or gravel, is not available at the site, it should be brought in. Backfill should be compacted in layers not exceeding one foot of thickness. This minimizes the tamping pressure necessary for compaction, thus decreasing the possibility of damage to the walls.

If absolutely no good backfill is available within hauling distance of the site, several horizontal drains may be necessary. There is another option, a product called Enkadrain®. Enkadrain® is a fibrous mat about ½" thick which is placed against the insulation (or protected membrane in the extreme south). Hydrostatic pressure is eliminated from the wall, because any water entering the tough nylon mesh is immediately carried down to the footing drain. Enkadrain® is not a substitute for a waterproof membrane; it is a complement. The 0.4" thickness is suitable for depths up to 10' below grade and was priced at about 50¢ per sq ft in September of 1991. In some cases, it may be cheaper than hauling in special backfill. For information, write: AKZO Industrial Systems Co., P.O. Box 7249, Asheville, NC 28802.

Another excellent product, Dow Styrofoam®

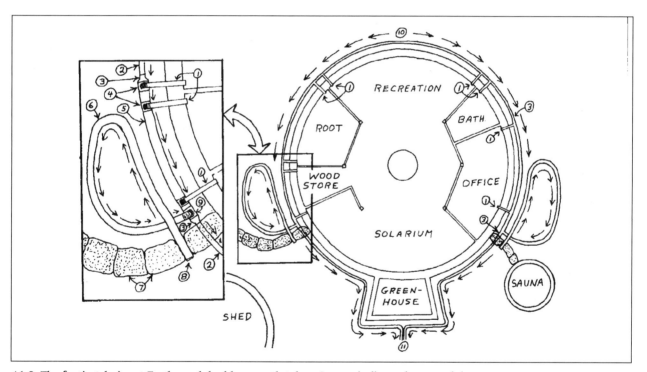

16-9. The footing drains at Earthwood double as earth tubes. Arrows indicate *downward* slope.

1. Earth-tube air outlets, made of 4" ABS plastic pipe.
2. 4" perforated drain tubing wrapped in filtration fabric.
3. 6" to 4" reducer.
4. 6" T-junction, with a 6" to 4" reducer and a 4" elbow above, as seen in 16-11.
5. 6" perforated tubing wrapped in filtration fabric.
6. 6" nonperforated flexible tubing, spiralling down under berm.

7. Retaining wall (made of large stones at Earthwood).
8. 6"-diameter earth-tube inlet. *See* 16-10. The inlet should be covered with a screen to stop entry by small animals.
9. 6" tee.
10. The high point of the footing drain system, near the north stake. The flow is in both directions from this point.
11. The footing drains change to nonperforated tubing at this

16-10. (left) The 6"-diameter earth-tube inlet is built into the stone retaining wall. It should be covered with a rodent-proof screen. 16-11. (right)The earth tubes at Earthwood connect with the footing drains.

Therma-Dry®, combines insulation and drainage in one application. An abundance of vertical grooves (21' deep by ⅜" wide) are cut into the outside of a 1½" (or 2¼") thick sheet of Dow Styrofoam® Blueboard, and covered with a nylon filtration mat. Water seeps into the grooves and is carried down to the footing drains. When one considers that insulation and drainage qualities are combined in a single product, the cost of $1.60 per sq ft for the 2¼" stuff is quite favorable.

At Earthwood, we relied on our footing drain and underfloor drain, as we were satisfied with the quality of our sand backfill, available right on site. The high point of the drain is at the north stake, and it slopes down gradually around the house in both directions, terminating in a large soakaway about 25' in front of the greenhouse. On a sloped site, the drain should flow to an exposed surface outlet, the opening of which is screened to prevent entry by insects or small animals.

THE EARTH TUBE

Figure 16-9 shows the route of our footing drains, which also double as earth tubes. Earth-tube inlets are placed in the corners of each room on the lower storey, as these are the locations prone to the collection of stagnant air. Note in the drawing that the air inlet tubes, located in the stone retaining walls (16-10), are 6" in diameter. This allows adequate air intake to feed four or five 4"-diameter earth-tube room outlets and increases the efficiency of the heat exchange within the tubes by presenting more surface area to the soil. Note that after two or three room outlets have been served, the pipe drops back to the ordinary 4" perforated variety, as seen in Illustration 16-11. The 6"-nonperforated earth-tube spirals down from the retaining-wall inlet, circles under the berm for about 25' or 30', and joins the footing drain from above by way of a T-junction. The footing drain continues on its slight downwards slope to the soakaway. No water can get trapped in the earth tube to close off the movement of air. Warm, moist air will condense on the inner surface of the earth tube and get carried away to the soakaway. This dehumidifies warm summer air as it cools it.

The perforated footing drains 'are bought wrapped in a nylon filtration sock to prevent clogging, and are surrounded by 4" to 6" of No. 2 crushed stone, as shown in 16-8. The important waterproofing detail where the earth tube enters the wall is carefully sealed with Bituthene®.

17

Backfilling

It is important to remember that several construction details proceed simultaneously: cordwood walls, block walls, masonry mass, post-and-beam, waterproofing, and drainage. We virtually finished the first storey prior to beginning the second. To do this, it was necessary to backfill the first storey to provide a place to stand for continued wall construction. The

walk-around deck provides a ready-made scaffold on the southern hemisphere. (See 17-1.)

Rigid-foam insulation is held against the Bituthene® membrane during backfilling, as shown in 17-2. This protects the membrane from damage during backfilling and later freeze-thaw cycles. Moreover, it renders the block wall useful as ther-

17-1. The walk-around deck provides a scaffold for second-storey construction.

17-2. Dennis and Jaki hold the Styrofoam® insulation against the wall during backfilling, while watching for large stones.

mal mass. We used 2" of Dow Styrofoam® on the entire bermed portion of the wall, except 3" down to frost level. The amount of insulation to use below grade will vary with the climate, right down to very little or none at all in warm climates. On straight walls, rigid foam insulation may be held in place by spot-gluing it with a mastic compatible with both the polystyrene and the membrane. No amount of mastic will hold rigid foam to a curved wall. During backfilling, we took care that no earth found its way between the Styrofoam® and the Bituthene®.

The backfilling should be compacted in runs of 12" or less. The bucket of the backhoe under the careful eye of Ed Garrow did most of this compaction for us, and we assisted with the heels of our boots. A hand tamper or a powered compactor would be of great use here, although the powered compactor may be

17-3. The Styrofoam® insulation exends to the top of the block wall. Chicken wire (l" mesh) is stapled to the first course of cordwood and tucked in behind the backfill or held to the Styrofoam® with small nails.

17-4. The bulk of the space is filled with cordwood masonry mortar, which is not expensive. A few days later, the mortar is covered with a 1/8" layer of surface-bonding cement, making a strong and fairly waterproof surface.

awkward to maneuver, especially if the backfill material is stony. Take care that large stones do not come crashing down against the insulation.

If the backfill has excellent percolation, a good-quality expanded polystyrene (also called bead-board or EPS) can be used instead of Styrofoam® at almost half the cost. Compare R-values when comparing costs.

We completed the second-storey construction by standing on the first storey backfilling. We roughly backfilled the upper storey in November with sand and gravel from the site. In the spring, after about 9"

of settling had occurred, we finished the berm with good soil, raked it smooth, and planted it with rye and wildflowers, with a mulch of old hay and straw to resist erosion.

We completed the transition from block wall to cordwood as shown in 17-3 and 17-4. Where backfilling is difficult, it is possible to apply surface-bonding cement directly to a sheet of well-scratched Styrofoam®. Note that the surface-bonded mortar detail carries water away from the cordwood masonry, and also that the detail is protected by a generous roof overhang of between 18" and 20".

18

Closing In

It was a literal race against time to get Earthwood closed in before winter. We hired, for six weeks, a second full-time man, Kork Smith, one of the most efficient workers I have ever witnessed. Kork kept Dennis, Jaki, and me well stocked with materials as we laid cordwood furiously through September and the first week of October, despite rain and unseasonably cold weather. We worked weekends in an effort to make up the lost time connected with the hardwood expansion debacle back in June and July. On October 9, we set our 2" × 6" wall plates on the cordwood wall. On the 10th, we installed the upstairs post-and-beam framework. On the 11th, we began installing the large five-by-ten and six-by-ten rafters. During the week of October 12 to 16, we finished the rafters, decked the roof, and applied Bituthene®. Bad weather limited us to four days the following week, but, luckily, we now had some important indoor work to do.

SNOWBLOCKS

There are 32 small panels of cordwood masonry to fill in between the plate system and the roof deck, separated from each other by the rafters. We call these the snowblocking panels, as they are comparable to the snowblocks installed between rafters on normal framed houses. The work is slow and tedious, because the mason is constantly having to fit the work to a rafter or to the ceiling… and in very cramped circumstances. The top insulation cavity in the cordwood masonry is much more easily filled with wads and strips of fibreglass than with loose sawdust insulation. The last mortar gap can be filled by pushing mortar off the back of a trowel with a

pointing knife, snugging the mud up against the packed fibreglass insulation. (*See* 3-23 and 3-33.) The snowblocks took us three days and the weather was cold and wet. Despite being under cover, the windows and door were not yet installed and the working conditions were miserable, especially for the person pointing the mortar on the exterior of the building. We were experiencing December weather in October.

DOORS

The following week, we installed the doors and windows. The front door was made by hand, 4" thick. I stayed with my old standby insulated door design which we'd used at Log End Cave and again at the Log End Sauna. (*See* 18-1.)

The downstairs door, which would eventually lead into a greenhouse, is a wood-framed, sliding-glass door unit, which must be installed according to the manufacturer's instructions. I strongly advise wood-framed units, as opposed to aluminum ones which conduct cold through the frames at an astonishing rate and result in frost buildup inside.

WINDOWS

We kept all our windows about 3" in from the outside of the building, which yields generous shelf space inside for plants. The two large windows in the living room have their wide sills at the perfect height for sitting, completing the conversation area. Installation of windows will depend on the kind that

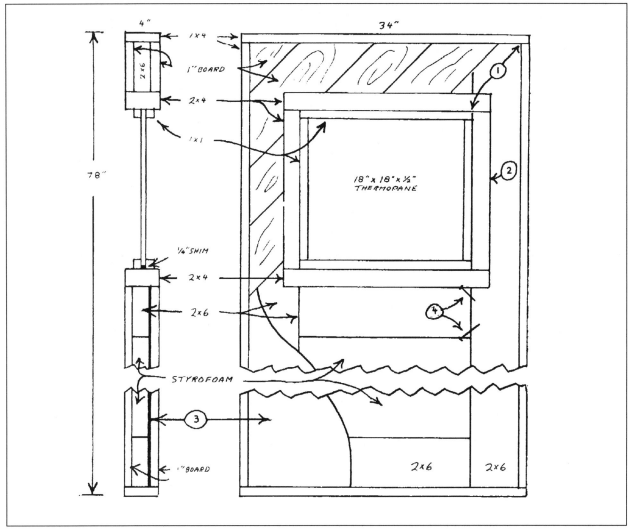

18-1. A cutaway view of the handmade door at Earthwood.

1. Silicone caulking at all through joints.
2. Two-by-six checked out for window framing.
3. 15-lb felt under outside sheathing.

4. Two-by-six framework held with toenails until 1" sheathing is fixed.

you have. With openable units, follow the manufacturer's instructions. With thermopanes, the units should rest on ¼" wooden shims, and a clearance of approximately ¼" should be maintained on the other three sides. We used rough-cut one-by-one molding on both sides of the unit, and caulked between the molding and the frame as well as between the molding and the glass. Any true silicone caulk is excellent, but rather expensive. A more economic alternative is a very good acrylic-and-silicone hybrid caulk manufactured by Red Devil, Cuprinol and others.

19

The Earth Roof

We have changed our roof design since the publication of *Underground Houses*. Although we have had no trouble with the Log End Cave roof, I feel that the design shown earlier in Figure 8-7 has several improvements. Firstly, the insulation is on the topside of the waterproof membrane, protecting the membrane from damage and freeze-thaw cycles. Secondly, we have begun to use a drainage layer of No. 2 crushed stone to help take a lot of the pressure off the membrane.

ROOF INSULATION

On the roof, we used 4" of Styrofoam®, an insulative value of about R-21. With the planking and earth values added, my guess is that the grand total of the roof insulation comes to about R-30. Remember, too, that new powdery snow has an R-value of about R-1 per inch of thickness.[11] And the earth roof, in my experience, holds snow better than any other kind of roof.

The installation of the Styrofoam® on the 16-faceted roof proved to be much easier than expected. I consulted a diagram which I'd originally drawn for Cordwood Masonry Houses and which is reproduced here. We used 2"-thick 4' × 8' sheets of Styrofoam®. Each of the 16 facets requires two full sheets for each course, cut as shown in 19-1, *Pattern One*, and one quarter of a sheet cut as shown in *Pattern Two*. Thus, a sheet cut according to *Pattern Two* will serve four facets, one course deep. Thirty-six sheets of insulation are required to cover the roof surface, one course thick. So 72 sheets are required

in all to give the full 4" cover. These patterns, which worked out perfectly for our outside wall diameter of 38' 8", produce virtually no waste. The patterns would be useful for house diameters of a foot different from ours, either way, but other patterns would have to be worked out for considerably larger or smaller diameters. (An admitted advantage of rectilinear earth roofs, of course, is that rigid-foam insulation—as well as sheet membranes—can be installed with much less cutting of material.)

The second course of Styrofoam® offsets the first course by 2", so that all of the radial laps are covered. In fact, we actually assembled full 4"-thick panels with this offset—or shiplap—built right in. (*See* 19-2.) Four to six 16-penny common nails tacked the panels together until we began to cover the roof, at which time they were removed. It is not necessary to buy tongue-and-groove Styrofoam® if this method is used. In fact, Styrofoam® T & G would not work out well on a multifaceted roof of this kind.

We used an inch of Styrofoam® on the overhang, right to the edge of the roof. Obviously, the overhang does not need to be insulated, but an inch of rigid polystyrene here—any kind—acts as a good protection for the membrane, especially where the railroad ties are used at the edge of the roof as retaining timbers. Any other places where joins between sheets are not shiplapped are also covered with an inch of scrap Styrofoam®.

At this point, cover the insulation with a single large piece of 6-mil polyethylene. This layer does not have to be absolutely watertight—that is not its purpose—but it will shed over 90 percent of the water before it gets to the membrane. This is cheap protec-

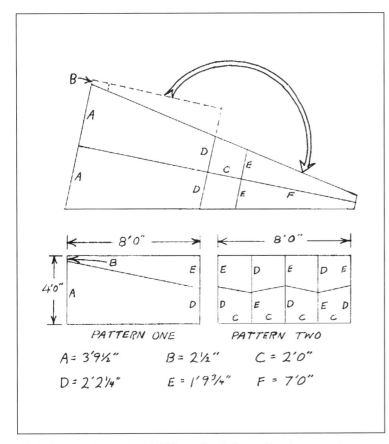

19-1. Patterns for cutting rigid foam insulation without any waste.

tion, and I cannot overemphasize its importance.

RETAINING TIMBERS

Something has to keep the crushed-stone drainage layer and the earth from falling off the roof. Two alternatives are eight-by-eight pressure-treated timbers (expensive) or old railroad ties. We chose the latter, as we had an opportunity to get 20 ties in excellent condition—*delivered*—for $100. A local railroad was closing a section of track nearby. The rough appearance of the ties is in keeping with the architecture of the building, although I must admit to being somewhat underwhelmed by the appearance of retaining walls constructed of old rail ties. The 8' 6" length of the ties was perfect for the outer edge of the roof, although the ends had to be cut at the facet angle so that one would abut cleanly to the next. The ties are truss-plated to each other all around the circle, but are not fastened

19-2 Placing rigid insulation on the roof.

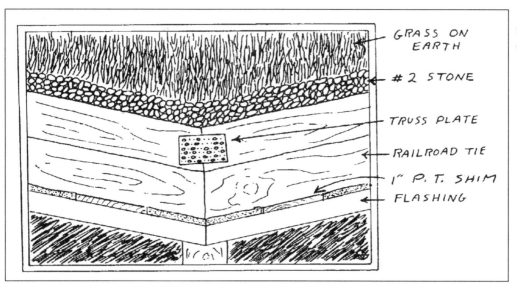

19-3. Retaining timbers are truss-plated to each other.

GRASS ON EARTH

#2 STONE

TRUSS PLATE

RAILROAD TIE

1" P.T. SHIM

FLASHING

to the roof by any other agent, save gravity.

As they weigh over 200 lbs each, gravity does a good job. (We placed these back-breakers onto the roof five at a time with the front bucket of a backhoe, which was on site backfilling the north wall.)

A few years after building the roof as described herein, we had our first and only leak. It had been raining for a couple of days, but the temperature was less than 32°F. Water would run through the drainage layer and then freeze when it got to the edge of the overhang, as it tried to get out between the railroad ties and the 1" Styrofoam® left on the overhang as a protection board. This situation is called an "ice dam." The water has no choice but to back up on the roof. We've already said that we can't dam the water out of the house. Drainage is the better part of waterproofing. When the drainage can't work, we're in trouble.

The water found its way through a small hole in the Bituthene® over the cordwood wall in the bedroom. It appears that the hole was caused by carpenter ants, which we could hear in the roof that winter. They sounded like mice. In the spring, we tore up a section of the sod, found and repaired the leak, eradicated the carpenter ants, and—this is important—shimmed up the railroad ties all around the house with 1" wooden shims, so that the chance of ice damming is greatly reduced. These shims can be cut from a plank of one-by-six pressure-treated lumber, and their location is shown in 19-3. We have had no trouble since their installation, despite some severe freezing conditions. But my view on retaining timbers has changed somewhat.

19-4. The sauna roof. The old barn timbers are truss-plated together, the same as the railroad ties on the main building.

MOSS SODS

Remembering that underground architect and planetary patriot Malcolm Wells always used moss around the edges of his earth-roofed buildings, I decided to try the same thing on my office. Fortunately, we had a plentiful supply of moss growing out in the sandy gravel pit in front of the house. With a flat spade, I cut it into sods about 5" high by 5" wide by 10" or 12" long and placed them like bricks around the edge of the office.

Figure 19-5 shows the sods after knitting together for a couple of years. I am very enthused about this moss business, and have used it on two other small buildings. It is natural, beautiful, and the drainage seems to be good. No leaks.

EARTH ROOFS: THE TRADE-OFFS

Beyond 10" of earth on the roof, the builder is involved in a very poor economic trade-off. The comparatively slight additional thermal advantage gained from, say, a 3' or 4' earth cover is greatly outweighed by the additional structural cost of the house.

In the case of owner-builders using plank-and-beam roofing, the structural members become absurd in their dimensions, and cranes become necessary for their installation, so the low cost and ease of construction advantages are lost.

Some of our seminar students, no doubt influenced by magazine articles showing 3'-thick earth roofs, are somewhat taken aback by my commentary on this. I have shaken one of their preconceived notions on earth sheltering; i.e., heavy roofs are a necessary part of the construction—the "massive passive" approach.

They are almost disappointed that they don't have to engineer for 450 lbs-per-sq-ft roof loads. In fact, some are so disappointed that they ask, almost defensively, "If less earth is better, why put any on at all? Why not simply build a well-insulated conventional roof engineered for the relatively small load consideration?"

A good question, and it deserves a good answer, so I whip out my chalk and scribble out the following list of advantages to be derived from 6" to 12" of earth:

(1) *Insulation:* Insulation may be the toughest concept of all to impart, so I'll treat it first. Some of our students arrive at the seminars still adhering to the popular misconception that the earth is a great insulator and, therefore, the more earth the better.

Early in the course, we explain that the advantage of earth sheltering is the setting of the structure into an ambient of higher temperature in winter, cooler in summer. If it's -20° F outside a house on the surface must be heated 90° to achieve comfort level. The low base temperature at Earthwood is about 30° F, so we have only 40° to go.

Having said that, I must add that earth is not completely without insulative value, but that by far the greatest value per inch is found in the first three

19-5. These moss sods drain well, retain the earth, and look natural.

or four inches. This is the part of the earth cover which is aerated by the root system of the green cover. The soil here is rich, loamy, and not greatly compacted.

And, finally, the grass cover itself, especially if allowed to grow long, supplies additional insulation. By comparison, the 4" of earth between 12" and 16" is compact, void of significant aeration by root systems, and usually of a non-humus nature. In short, that layer of earth has only a fraction of the insulative value found in the top 4". Stones, which are deliberately kept out of a shallow roof, are tremendous conductors of heat.

Obviously, the correct amount of rigid insulation must be applied according to the climate, but the earth, especially the first few inches, does supply an additional value. Increased insulative value, however, is more economically achieved with more rigid foam, not more earth, with its accompanying support structure.

A related note on insulation: because of higher surface temperatures due to solar energy on a composition roof, less snow is retained. The earth roof, in my experience, holds a thicker snow cover, adding greatly to the insulation when compared to a "traditional" roof.

(2) *Drainage:* The conventional roof does not slow runoff. During a "frog strangler," a tremendous amount of water is added to the earth next to the house, increasing hydrostatic pressure and taxing the waterproofing and drain systems.

This can be eliminated by gutters, an additional cost, and the roof must be kept high enough off of the ground for their installation. And, that water still has to be carried somewhere. We have found that with a freestanding 9" earth roof there is much less runoff during a rainstorm. Sometimes, the earth roof may be completely saturated and will not hold any more water, but these conditions occur infrequently in areas of moderate rainfall. The earth roof moderates runoff. Much of the water on a shallow roof is lost back to the atmosphere by evaporation or used by the green cover.

(3) *Esthetics:* One of the big plusses of earth-sheltered housing is that the home harmonizes with the site. Ideally, the natural contours of the land are retained. The earth roof, in my opinion, is much more in tune with this concept than even the best asphalt, wood, or fibreglass shingles.

(4) *Cooling:* The evaporative effect on a sod roof

aids in summer cooling. Even 6" of earth protects the substrate from the high surface temperatures found on a composition roof.

(5) *Longevity:* Properly built, the earth roof is the longest-lasting. Other roofs are subject to deterioration by the sun's ultraviolet rays and have a limited life span. The best available shingles are estimated to a 35- to 40-year life span. Even a 6" sod roof is sufficient in protecting the membrane and insulation from breakdown by ultraviolet.

(6) *Ecology:* The earth roof supports life. Here's a chance to save one or two thousand square feet of the planet's surface from being converted to a lifeless desert. Wild flowers and leafy vegetables are possible on a shallow roof, as well as lawn area. The roof is not the right place to grow shrubs and trees.

In combination with proper protective fencing, the roof can extend the yard space. This is of particular importance on small lots.

(7) *Protection:* Admittedly, the thicker earth roof is a better protection against sound and radiation than a shallow roof, but again, as with insulation, the first few inches have the greatest impact on these considerations.

The primary gain from putting 3' of earth on the roof as opposed to 1' is that the entire structure is effectively set 2' deeper into the ground. At soil depths in the range of 6' to 12' there is an average difference of $1°F$ for each foot of depth, both in the coldest month (March) and the warmest (September).

In other words, setting the structure 2' deeper amounts to a $2°F$ gain, both for summer cooling and winter heating, plus the slightly greater insulation derived from the extra 2' of earth cover.

It is my contention that an extra inch of rigid foam insulation is a much more cost-effective means of achieving the same advantage than building a structure capable of supporting a 3' earth load.

The primary concern in choosing soil thickness should be the maintenance of the green cover. We have been able to maintain rye grass on our 6" earth roof for nine years without watering. And the grass continues to improve. I ascribe this to our use of soil with poor drainage characteristics: silt drawn from the bottom of a drained pond. In areas without much rain, I would recommend 10" to 12" of earth to support a deeper root system and use of a local grass exhibiting good ability to bounce back from drought.

The Drainage Layer

Remember: drainage first, then waterproofing. On top of the hay or straw protection layer, we placed about 2" of No. 2 crushed stone. When the earth becomes fully saturated, this drainage layer will direct the excess water to the edge of the roof, taking most of the workload off the waterproofing membrane. A filtration mat of hay or straw under the rail ties filters the runoff. Another 2" to 3" of hay or straw filtration mat above the crushed stone layer prevents the topsoil from clogging the drainage layer. These filtration layers compress greatly under the weight of the earth and do not add appreciably to the thickness of the roof. Look for cheap bales which, for some reason, are not valuable as animal feed. Actually, straw is preferable to hay, because of the straw's tougher fibres.

We brought the crushed stone onto the roof with the front bucket, too, and spread the stone by hand using shovels and 5-gallon plastic buckets. Crushed stone should also be placed up against the rail ties, to eliminate outward hydrostatic and frost pressures. It took four of us about 2½ hours to cover the roof with stone, and our arms were several inches longer at the end of the job.

The Earth

We did not put earth on the roof that fall. It was too late for planting, and we would not be moving into the house until spring, anyway, so the little extra insulative gain was not important. In August, 1982, while the backhoe was again on site for finish landscaping, we placed the earth. (See 19-6 and 19-7.)

The "topsoil" we chose was actually a silty mixture which had been scraped from a stream's flood plain down in the valley below. The soil is stone-free and has a relatively poor percolation, which is good in this instance as a lesser thickness of earth is required to retain water for use by the plant cover. We used about 8" of earth. The drainage layer takes care of oversaturation.

We planted rye in early September and, thanks to a beautiful autumn, had grass coming through the thin mulch layer within weeks. By the end of October the grass was thick, green, and lush.

A roof built up as described should last forever.

19-6. (left) Perched on the north berm, the front bucket of a backhoe deposits the topsoil on the roof. 19-7. (right) Jaki spreads the topsoil with 5-gallon buckets. No, she didn't do the whole roof by herself. Somebody has to take these pictures, you know.

20

Electrical and Plumbing Systems

It is not within the scope of this book to describe wind energy or photovoltaic electrical production. (Photovoltaic cells convert sunlight directly into electricity. They are now less expensive than wind systems and require less maintenance.) There are many excellent books, with more appearing all the time, as the technologies on these subjects improve. What is pertinent here, however, is a discussion of the use of 12-volt DC power as an alternative to standard 115-volt AC house current.

THE 12-VOLT ELECTRICAL SYSTEM

Our Sencenbaugh 1000-watt wind plant, which is mounted on a 110' Rohn guyed tower, delivers power to a control box. This control box serves several functions: (1) It regulates the power which is delivered to the battery bank, acting as a voltage regulator. When the batteries are fully charged, additional power from the wind plant is converted to heat and dissipated. (2) It monitors power consumption in the home, that is, the draw on the batteries. A meter measures this draw in amperes. (3) Another meter shows the output of the wind plant, also in amps. A *low* or *no output* reading in high winds would indicate a problem somewhere else in the wind system. Similar control boxes are used in combination with banks of photovoltaic cells.

Two wires join the control box to the battery bank. At Earthwood, this used to be composed of 14 6-volt, deep-cycle storage batteries, each with a rated capacity of 200 amp-hours. The batteries are con-

nected in both series and parallel, as shown in 20-1 and 20-2. First, pairs of two batteries are connected in series to provide a 12-volt unit. Then, the seven 12-volt units are connected in parallel to accumulate the power storage while keeping the voltage at 12. Each pair of batteries, then, stores 200 amp-hours at 12-volts. The total storage capacity of the battery bank is 1,400 amp-hours at 12 volts.

BATTERY UPDATE

Our 14 Keystone batteries lasted until the fall of 1990, which means we got about 8½ years out of them, which I consider to be very good service indeed. We were careful to keep them topped up with distilled water. One battery with a bad cell had to be replaced after about 4 years. During the last year, the batteries were getting quite tired. We took two weak ones off-line, which was okay as we now had photovoltaic back-up to our wind system and didn't have to go through as many days without energy imput. A bad battery, or even one bad cell, will bring the voltage of the whole bank down dramatically.

We went through about six months of sparse times, energy-wise, when the wind plant was off-line with mechanical difficulties and the batteries were failing. We replaced the batteries with 10 Exide GC-4 batteries, still the 6-volt deep cycle variety. We presently use six Arco M-75 photovoltaic panels, each with a maximum output of 48 watts when the sun hits them at perfect right angles. With the wind

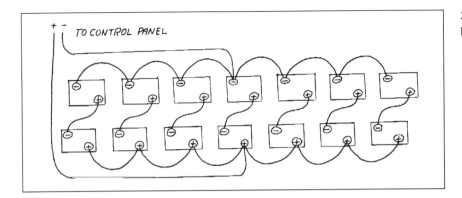

20-1. Batteries are connected in both series and parallel, as shown.

20-2. The battery storage system consists of 14 six-volt deep cycle batteries. The pedal-powered water pump can be seen at the right of the picture.

plant back on line, we now seem to make power to spare. The ten batteries are enough.

It is easer, I think, to calculate in terms of *watts*, an empiric unit of electrical power. This eliminates the confusion which can arise from systems of different voltages: 6-volt, 12-volt, 32-volt, or 115-volt. Amperes times volts equals watts, so the 1000 amp-hour storage capacity at 12 volts can be legitimately thought of as a storage of 12,000 watt-hours. (For the health and long life of the batteries, actual discharge should not exceed about half of this rated capacity.) But what does this figure mean? Well, a 50-watt light bulb consumes 50 watts for each hour that it is lit. If the bulb burned constantly for 100 hours—or two 25-watt bulbs for 100 hours each—the total draw on the batteries would be 5000 watt-hours.

At first, these figures may not sound very impres-

sive. It may seem to the reader that the reserve capacity of the system is not very great. Well, while it is undeniable that personal consumption patterns must, by necessity, be lower with a battery storage system than those (which are disgraceful, from a planetary viewpoint) in the average home, there are a number of other factors which come into play in our favor. For one thing, an incandescent 12-volt bulb delivers about twice the lumens per watt as ordinary 115-volt AC bulbs. (The lumen, like candle-power, is a unit of measuring light intensity.) So we are able to use much lower-wattage bulbs, 25 watts in most cases.

Fluorescent lights can further increase the efficiency by a factor of 2.5, but it is important to choose high-quality 12-volt fluorescent fixtures. Jaki and I have had bad luck buying cheap fixtures from recreational vehicle suppliers. The bulbs fail at low voltage and are expensive to replace. The higher-quality units are considerably more expensive ($30 to $50 on average), but have an expected life of about 7500 hours. Also, replacement bulbs for the better units are the same as used with 115-volt AC power, so they are not expensive and are available anywhere.

PHOTOVOLTAIC (PV) CELLS

Photovoltaic cells take in light (the *photo* part) and—almost magically—convert the light energy into 12-volt DC electricity (the *voltaic* part.) They can be joined in different combinations to make (most commonly) 12-volt and 24-volt systems. Besides low maintenance and high reliability, one of their best features is that they are "modular," which means that you can add on

20-3. The PV panels at Earthwood face due south.

20-4 Power from the son! *(Suzanne Moore photo.)*

to your system as finances allow. A small basic kit, for example, was available from Fowler Solar Electric, Inc. in late 1991 for $960. The kit consists of two Hoxan 48-watt PV panels, two Trojan T-105 batteries and a charge controller kit. This kit would supply power for 12-volt low wattage lighting, a small TV and occasional use of other small appliances. And it can be expanded.

Our six 48-watt panels complement the wind plant beautifully. There is plenty of wind in the winter and plenty of sun in the summer, so we never go very long without power input. Our ten batteries supply the electrical storage capacity to get us through several calm overcast days. But for someone starting out from scratch today, I would still advise going straight solar and leaving wind energy behind. The exception would be if you have a remote site with lots of wind, and can get a great deal on a good used or reconditioned wind plant.

SON POWER

Our son Rohan, who is a keen mountain biker, is our back-up power source. In 1991, we bought a bicycle generator with a stand suitable for mounting an ordinary multi-speed bike. Rohan can put out 100 watts for short periods of time—a few minutes—but can maintain outputs in excess of 50 watts for a half hour or so. This is equivalent to a PV panel at peak output and provides good winter training. We bought the unit from Real Goods, listed with other alternative energy suppliers below.

INVERTERS

In 1988 we added an inverter to our home power system. This device "inverts" low voltage DC current to standard 115-volt AC house current. We use it for a small color TV, vacuum cleaner, blender, VCR, power tools, a small washing machine, and a slide projector. We still use 12-volt lighting, although many experts in the field now recommend using an inverter for lighting as well. With 115-volt power, ordinary house wiring can be employed, and there are now some very efficient and relatively low-cost bulbs on the market (compared with 12-volt lights.)

PRACTICALITY VS. IDEALISM

We love being unplugged from the power company at Earthwood. There is a quiet joy that comes from being free of utility and fuel bills (except for telephone and about $250 a year for propane gas.) But the reader should be aware that homemade power is very much more expensive on a per-kilowatt-hour basis than is commercial power. If the power lines come right by your site, and there is no exorbitant hook-up charge, the reality is that it is cheaper for you to plug in. If you want to stay unplugged for spiritual or philosophical reasons, great. But don't do it for purely economic reasons unless the cost of bringing in power would exceed three or four thousand dollars.

Our Mushwood cottage is only used for a few weeks in the year. The site has poor wind and solar potential because of high fir trees. The commercial power is right there. The logical decision—and the one we made in late 1991—was to plug in. We know how to live well with very little electrical energy, so we do not expect big electric bills. And if we want to rent out the place, we can. We rented out Log End Cave one winter and the tenants—unused to "gentling down" to lower patterns of consumption—fried our batteries in short order.

ALTERNATIVE ENERGY SOURCES

These dealers carry a full line of solar electric equipment including photovoltaic cells and all the components that go along with them. Most also carry wind and hydro equipment, deep-cycle batteries, inverters, back-up generators, 12-volt lights and appliances, and other devices which diminish our reliance on fossil fuels and nuclear power. Nuclear power is a great source of free energy if located the right distance from your home (Earth): 93,000,000 miles.

Fowler Solar Electric Inc.
P.O. Box 435,
13 Bashan Hill Road
Worthington, MA 01098   413/238-5974

Real Goods
966 Mazzoni Street
Ukiah, CA 95482   707/468-0301

Backwoods Solar Electric Systems
8530 Rapid Lightning Creek Road
Sandpoint, ID 83864   208/263-4290

Talmage Engineering
P.O.Box 497A, Beachwood Road
Kennebunkport, ME 04046   207/967-5945

Home Power Magazine
P.O. Box 130
Hornbrook, CA 96044-0130

I've had dealings with all of the sources listed above, and do not hesitate in recommending them. People in this industry seem to be conscientious and honest.

WIRING

If a 115-volt circuit is to be used in the house, this should be wired in the normal fashion for household wiring. A 12-volt circuit, on the other hand, is a little different. The primary differences are threefold:

(1) All switches should be the positive click type. It will probably be necessary to special-order your switches from an electrical supply house. Specify "AC/DC" switches.

(2) Fuses, likewise, must be specially made for 12-volts. Ordinary screw-in AC fuses may fail to work properly in an overload situation. The 12-volt direct current will arc across the fuse.

(3) Low voltage requires heavy wire, or unacceptable line loss will result. I recommend the use of No. 8 copper wire for the main circuits, with No. 12 to the individual lights or receptacles. All our receptacles are wired for 12-volts. We have converted the cigarette-lighter-type plug on our appliances to 3-pronged plugs. The use of 3-pronged plugs is imperative, as many appliances, such as a 12-volt TV, will burn out if electrical polarity is wrong. Once correct polarity is established, make the plug conversion carefully and the third prong will assure that the appliance cannot be plugged in wrong. The connection of the ground wire is optional.

With a structure such as Earthwood, with its cordwood walls and heavy exposed timber framework, I strongly recommend the use of exposed wire mold, or rigid EMT conduit, which can be painted. The modern-day wire mold is attractive, safe, and easy to install. And easy to alter, should a change be desired. This wire mold will meet electrical codes, although a home-producer independent of the power company will not be overly concerned with this, especially as many code-approved components will not work with a 12-volt system. His primary concerns will be safety and convenience. Other options are to run the wiring inside internal walls or within a conduit built into the insulated mortar joint of the cordwood, but this is an awful lot of trouble for little gain when compared with good-quality surface-mounted wire mold.

PLUMBING

Again, I will concentrate on the differences between our system and standard plumbing practice. Not the least important part of our system is a simple old-fashioned hand pump located at the top of our well. If all else fails, we can take 5-gallon pails out to the well and draw water. I'm sure that many readers have experienced situations where power failures or some other problem with the plumbing or electrical systems have made water totally unavailable to the household. Wished you'd had a hand pump then, didn't you? In addition, we use the hand pump all the time for outdoor purposes, such as filling 55-gallon drums for mortar mixing, cleaning maple sap buckets, and washing cars or dogs. In a cold climate like ours, choose a pump that lets the water down into the cylinder a few feet after pumping stops. This prevents frozen pumps.

THE BICYCLE PUMP SYSTEM

If George Barber did not exist, it would be necessary to create him. This gentle man, creator of an amazingly simple and energy-efficient geodesic dome, master of the wood-fired underslab heating system, and surveyor extraordinaire, has added the pedal-powered plumbing system to his repertoire. I approached George with our goal of being able to draw water up from the well, the suction being provided by a pump, and then to push it upstairs to a cold-water holding tank. In response, George designed the system described below, with the driving force for the pump coming from pedal power. The water-consuming appliances would be either gravity- or pressure-operated. Water would be heat-

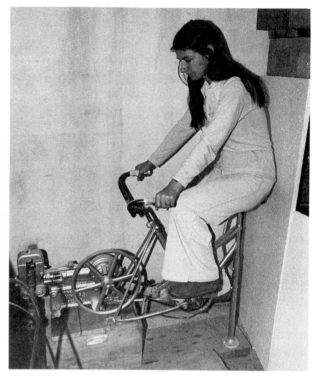

20-5. Jaki Roy demonstrates the "Waterwood" pumping system, designed by George Barber.

ed with wood in winter and propane gas in summer.

The first job was to gather together suitable components. We selected an old Myers double-action piston pump, rebuilt by an old-timer in town who has been working with such pumps all his life. New pumps of a similar kind are still available. For our cold-water reservoir, we used an 80-gallon hot-water tank whose electrical elements were burned out. This was given to us by the old-timer who sold us the pump. The hot-water tank is a new 30-gallon propane gas model, very well insulated for energy conservation. The bicycle frame was an old one which George had lying around the house.

One of the more critical jobs in the whole project was to fit the bicycle frame and the pump to a common rigid framework. Everything had to fit perfectly. Our bicycle pump would be located in the utility room, where the incoming line from the well enters the house. Using leftover ends of beams, I built a heavy wooden framework so that the suction end of the pump would be elevated to the same height as the 1¼" plastic line coming from the well, about 21".

George reversed the seat and handlebars on the old bike frame and fitted the rear-wheel support frame so that it straddled the pulley wheel of the Myers pump. He welded the threaded wheel-sprocket assembly from a bicycle right to the inner side of the pulley wheel. Now it was possible, by removing the wheel, to screw on different sprockets with various numbers of teeth.

By experimentation, we found that the best configuration was to use a sprocket with 16 teeth. The main driving sprocket, which the pedals were attached to, has 56 teeth. This gives us a gear ratio of 3½ to 1. One complete revolution of the pedals causes the pulley wheel of the pump to turn 3½ times. With a double-action piston pump, the piston moves in and out on each revolution of the pulley, pumping water on each stroke. Slow pedalling pumps about 7 gallons of water per minute, while 10 gallons and more is possible at higher speeds.

The discharge end of the pump feeds into a pipe which runs under the concrete floor and then rises into the upper storey where it feeds the kitchen-sink cold-water tap and continues on to the 80-gallon holding tank.

An overflow pipe comes out of the top of the reservoir (20-6) and discharges into the bathtub. There is a gate valve on this overflow line which can be closed if a pressure system is desired instead of a gravity system. This is the way we generally run our plumbing. In the pressure mode, the incoming water compresses the column of air at the top of the tank, building up good house pressure. With pedalling alone, it would be possible to build up a pressure great enough to burst the pipes, so we have installed a 100-lb pressure relief valve right near the discharge end of the pump.

The plastic float showing in the section of clear milking tube (20-6) indicates the water level in the 80-gallon holding tank. Side entrances to the tank were afforded by removal of the old electric heater element of this tank. A conversion fitting can be bought at plumbing supply houses to get to ¾" pipe. The spigot at the bottom of the tank allows the tank to be cleaned. The gate valve on the incoming line can be closed for working on appliances without draining the tank.

Note the juxtaposition of the hot-water tank with the cold-water reservoir. When water in the large tank drops below the horizontal pipe leading to the hot-water tank, the suction is broken and no

20-6. The upper tank is an 80-gallon cold-water storage, which can be pressurized by the pedal-powered pump. The lower tank is a 30-gallon propane-gas hot-water heater. The water in this tank can also be heated by the wood stove in the kitchen. The clear plastic milking tube with the float in it monitors the level of the large tank.

further flow is possible to the hot-water tank. Moreover, because the outlet of the hot-water tank is on the top, no more water can be drawn out of the tank. This is an important safety feature of the system and en-sures that the tank cannot be boiled dry; nor can an excessive steam buildup occur. In addition, there is a pressure- and temperature-sensitive relief valve at the top of the hot-water tank. Our Oval cookstove heats our hot water in the wintertime by means of a water jacket (also called a *water front*) on the side of the firebox. It's the old thermal siphon again. Heated water rises out of the top of the water jacket and enters the hot-water tank at an inlet near the top. The cold-water return comes from an outlet near the bottom of the tank, which formerly held a boiler valve for cleaning the tank. We transferred the boiler valve to a point in the line just below the water jacket, enabling us to drain the tank, if desired for any reason.

We shut off the gas water heater's pilot flame from December until April. As long as the cookstove is in use at least one hour a day, we have plenty of hot water. We used ¾" pipe to join the water jacket to the hot water tank. This connection is best made with copper rather than plastic pipe.

I have to say that the pedal-powered water system is one of the most successful innovations to come out of the whole Earthwood project, and one which was not even considered at the design stage, although we did install the underfloor line and the incoming line from the well with the idea that some sort of pump would be located in the utility room. Jaki and I find that 5 or 6 minutes of pedalling per day meets our water needs of 50 to 60 gallons. The exercise is pleasant, good for the heart, and helps keep us trim. We have hot and cold running water throughout the house, flush toilets, bath, shower, and hot tub... all without any form of electrical device. For this, we thank George Barber for his detailed design and precision craftsmanship, and Peter Allen for his kind assistance.

21

Interior Walls and Finishing

During the winter of 1981-82, we worked under cover at finishing jobs that didn't require a great deal of money. Thanks to the hardwood expansion debacle, our checking account was rather hollow. Luckily, we had a rent- and mortgage-free house—Log End Cave—to live in. An occasional firing of the masonry stove kept Earthwood at a comfortable 50° F to 55° F. working temperature.

INTERNAL WALLS

Framing for all internal walls is standard two-by-four (actually 1½" × 3½") construction, 16" on center. Internal walls always rise up to meet the underside of the five-by-ten roof beams. We had plenty of old 1"-thick hemlock boards of varying widths to cover the lower portion of the walls, up to a height of 4'. The boards were by-products of sawing the four-by-eight floor joists three years earlier, and eliminated the need for bottom molding or skirting boards. Above the rough-cut boards, we switched to ½" drywall (or plasterboard), so that we could create bright, light-reflecting walls at low cost. Drywall is inexpensive and makes a good-quality wall when used with 16" o.c. framing.

In some places—such as the bathroom, the kitchen, Rohan's bedroom, and behind our own bed—we simply painted the drywall, expertly fitted and taped by our friend Ron Light. At other locations, Jaki applied a textured paint of our own making, using Ron's nylon-textured roller. This came out very nicely, introducing a pleasing wall texture to the liv-

ing area, the stairwell, and on some of our bedroom wall panels. We make a low-cost textured paint by combining premixed drywall joint compound with the cheapest white latex paint, at a proportion of about 4 to 1. The latex paint brightens the mixture and gives it the right consistency for rolling. This mix is only a fraction of the cost of commercially available textured paints and is every bit as good.

UPSTAIRS FLOOR: TONGUE-AND-GROOVE SPRUCE

I borrowed a floor sander from a friend and sanded and sanded… probably for about six days, although it seemed much longer at the time. Instead of varnish and polyurethane, which must be completely recoated every few years, we decided to go with an oil-based floor finish which penetrates into the wood while helping to harden it. If a wear pattern develops at high traffic locations, it is an easy matter to feather in additional oil without the need for complete floor refinishing.

THE SLATE FLOOR

We are very proud of our low-cost slate floor, the central portion of which can be seen in 21-1. The slates are recycled roofing tiles, turned with their weathered side down. I traded a Ping-Pong table to Jonathan Cross for half of the slates; the rest were

211

21-1. The slate floor draws the eye to the masonry stove at the center of the building.

free, given to me by a fellow who was renovating an old house nearby. Here's what we did:

(1) I mixed a thin slurry of sand, Portland cement and pure Acryl-60 bonding agent (made by Thoro System Products, 7800 N.W. 38th St., Miami, FL 33166.) With a lamb's-wool roller, I applied the slurry to the smooth concrete floor and to the side of the slates which were to be hidden. This was the weathered side, remember, and it had some very nasty sharp edges which could, if left exposed, cut a stubbed bare foot or toe. The mixture dries in a half hour.

(2) I applied a ⅜" bed of mud to the section of floor where I was working, using ⅜" thick wooden laths as depth guides for screeding. The mix was: 5 screened sand, 1 Portland cement, 1 masonry cement.

(3) The slates were placed into this bed of mortar, using a small rubber hammer to set them in all at the same level. I used a rented slate cutter to make slates fit in certain places, such as near the center, but most of the slates were laid without the need for cutting.

(4) We pointed between the slates with a pointing knife. Chief pointer Jaki did most of this work, but the whole family chipped in on this beautiful but labor-intensive floor. We had done about 40 percent of the floor several years ago, but finished it in six long days in 1991. On our best day, with Rohan, Jaki,

and I all working, we laid about 80 of the 10" by 14" slates. Thanks to the Acryl-60 based slurry, the bond is tremendous. You can't pull the slates up with a crowbar. I've tried.

(5) After about 10 days of curing, we applied a slate sealer to bring out the color.

The slate tiles cover all of the downstairs living space, but not the bedrooms, bath, or utility room. Rohan's bedroom is carpeted, Darin's has a painted floor with throw rugs. The bathroom floor is vinyl.

KITCHEN CABINETS

The curved walls really do not present a problem with regard to decorating, because cordwood masonry comes already decorated and the surface-bonded block walls are simply painted. (You can't even see the blocks.) And, interior walls are all straight and flat, so any kind of surface can be used. The only time that the curved walls slowed us down was during construction of the kitchen cabinets, which we fitted to the external wall. This custom-fitting probably cost us an extra 25 percent in both hired labor and materials, but the total cost of the quite expansive cabinets—eleven large cupboards, a four-drawer unit (bought new for $109), 56 sq ft of

21-2. The cabinet doors are made of quality pine-faced plywood.

laminate countertop, a stainless-steel sink unit, and display shelves—was about $650.

LOW-COST CABINET DOORS

One of the main reasons that we came out of the kitchen cabinets at a relatively low cost was that Jaki and I made our own inexpensive cabinet doors and surface-mounted them to the framework crafted by friend Bob Smaldone. (Yes, a lot of friends got their hands on the Earthwood project. Whenever you have the opportunity to enlist the help of someone who can do the job twice as well and twice as fast as you can yourself, take it... especially if you can trade off the time on one of his projects, or barter the labor in some other way.)

The doors were simple to make and hang. We bought a 4' × 8' sheet of the highest-quality ¾" plywood, pine-finished on both sides to match the one-by-six pine cabinet material. This sheet of plywood cost $46. That doesn't sound so cheap, you may be thinking. But wait. With a friend's table saw, we cut the plywood into 12 doors, each measuring 16" × 24". This took about ten minutes, as there are actually very few cuts to make and absolutely zero wastage of material (except for the saw kerfs themselves). All that remained was to sand

and oil the doors. So each of the 12 doors cost us about $4, and, if you don't think that's cheap, you haven't priced cabinet doors lately. By surface-mounting them instead of flush-mounting, fancy fitting was avoided and labor time and cost were lessened. The result is quite attractive (21-2).

The information in the previous paragraph could save you ten times the cost of this book, but don't try to gild the lily by buying cheap plywood. Warping and dissatisfaction will be the likely results if you don't get the best-quality stuff.

RUSTICITY EQUALS LOW-COST COMFORT

Note that the commentary on interior finishing has been quite short, whereas in a "normal" house, the finish work constitutes a substantial part of the time and money expended. This is where the rustic style of cordwood masonry, stonework, and exposed heavy timbers and planking really pays dividends. Most "finish" work is completed during construction, and the result is a warm and cozy atmosphere. I won't even waste a new paragraph, much less a chapter, on exterior finishing, which consisted of painting the window frames of the openable units.

22

The Solar Greenhouse

The greenhouse really designed itself, based on: (1) our goals for the space, (2) the architecture of the house, and (3) available—largely leftover—materials.

GOALS

The greenhouse should provide:

(1) a frost-proof place to start seedlings for the raised-bed garden, in order to get a head start on the short North Country growing season.

(2) an opportunity to grow some hardy vegetables, such as Chinese cabbage, throughout the winter.

(3) a second entrance to Earthwood, with a thermal break as entry and egress are made.

(4) a secondary source of heating the house on sunny winter days.

(5) a sun room for wintertime lounging.

(6) a support structure for the second-storey deck.

(7) a pleasing harmony to the esthetics of the house.

ARCHITECTURAL CONSIDERATIONS

The unifying architectural themes of the house construction are the heavy post-and-beam frame, the radial rafter system created by the circular shape, and the cordwood masonry.

MATERIALS

The greenhouse had to be inexpensive, by necessity. A perusal of the pile of leftover timbers revealed that we had one rather attractive hand-hewn six-by-six beam, which we had considered to be too small to be in harmony with the other internal posts in the house, although it was certainly strong enough. Cut in half, the 10' beam yielded a matching pair of 5' posts. As these were kind of short for posts, we built a 17"-high cordwood knee wall around the perimeter of the greenhouse, leaving a place for the door, which faces east towards the sauna. Now the posts would be the right height to support a six-by-ten girder on the south side of the greenhouse. The girder was a spare roof rafter, supporting the outer ends of the main four-by-eight greenhouse rafters.

The shape of the greenhouse is trapezoidal, so that the main roof-support rafters would radiate outwards from the house as an extension of the first floor joists. In fact, we'd left the south floor joist extending out about 10' during the house construction for this very purpose, although the rafters immediately to the east and west of the south rafter had to be extended during construction of the greenhouse.

We used the built-up corners (or *stackwall*) method of cordwood construction. Leftover pieces of four-by-eight floor joists laid side by side in the corners established the 16" thickness of our cordwood knee wall. By keeping the glass to the very outside, a substantial and very useful shelf for starting plants is created on the interior. The cordwood was capped

22-1. The roof structure of the greenhouse supports a deck above. The original greenhouse was just half its present size.

with three two-by-six planks, forming the shelves. Note that the corners are not true right angles, because the south side of the greenhouse is the base of the trapezoid. Shaping the corner blocks and the two-by-six planks with a chainsaw was necessary to obtain the proper angle, but this shows the flexibility of cordwood masonry.

Another trip to our local manufacturer of insulated glass yielded positive results. It happened that the firm had a good selection of "shop units" on hand. I'd made drawings of the east, west, and south openings formed by the knee wall and the heavy frame. At the plant office, I spent time going through the list of shop units and juggled dimensions until I figured out a way to fill the openings as neatly as possible. Two of the units were short of the required height by a few inches, but we took advantage of this by installing high- and low-screened vents at these places. The vent openings can be plugged with an insulated cover for the winter. All the window units

were 1" thick thermopane: two pieces of ¼" plate separated by a ½" dry air space.

THE ROOF

Prior to nailing the deck on the second storey, we had to install a "subroof" over the greenhouse to draw away water which falls between the decking. We chose corrugated fibreglass for the subroof and supported it by a network of two-by-four purlins (small cross-rafters), 16" on center. The purlins were set below the top level of the four-by-eight main rafters, with a slope of 2" in 10' established by the placement of the purlins themselves. The highest pair of purlins are closest to the house, with each succeeding pair ¼" lower to provide good water runoff. (See 22-2.) The corrugated fibreglass was cut to fit the unusual shape and we sealed the edges

22-2. The rafters, purlins, and corrugated fibreglass panels make up the subroof of the greenhouse. Planking for the deck is nailed directly to the rafters.

with—what else?—Bituthene®. (I have even repaired rips in my black rubber galoshes with that remarkable stuff.) We carefully nailed the decking above to immediately protect the weak fibreglass panels.

EXPANSION

The original greenhouse, the size of which can be seen in 22-2, was only two rafter bays wide and had glass on all three exposed sides. It proved to be too small and temperature fluctuations were too great. Finally, the sitting deck overhead was a little cramped for space. We cured all this by expanding the greenhouse as I'd suggested in the following plan which appeared in the original *Earthwood* book. (*See* 22-3.)

The only change we made was a slightly different placement of the insulated door, closer to the outside corner. This change has been a great success. With massive cordwood walls now on the east and west sides, and 2" of extruded polystyrene in the ceiling, the greenhouse, now more truly a solar room of 144 sq ft, holds its heat better. With the sliding glass door open in February, the solar gain helps to heat the house. On cold spring nights, the door can again be left open so that some of the house heat can prevent tender young plants from freezing. The solar room now meets the goals which we'd set out, and converts the heat loss of the large sliding glass door unit into a positive gain.

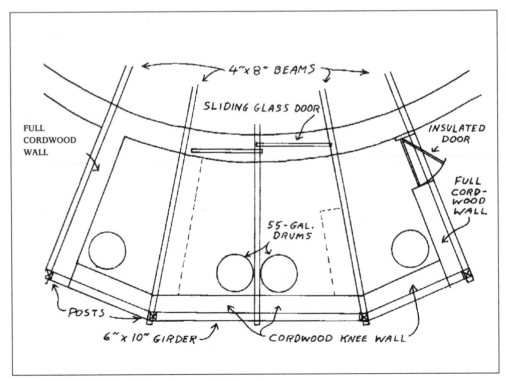

22-3. A better greenhouse design. The original greenhouse at Earthwood consisted of the middle two sections only.

23

The Cordwood Hot Tub*

One dip in a friend's hot tub was all it took for Jaki, Rohan, and me to become hooked on tubbing. Now, I won't take the time here to try to convince the reader of the sense of well-being, the health advantages, the spiritual calm, and all the other good stuff that hot tubbing provides. I'll leave that to other enthusiasts. Besides, one soothing dip in a good hot tub will be far more convincing than any amount of eloquent erudition on the subject.

We wanted a tub that would be cheap to build, cheap to run (energy-efficient) and in harmony with the building. So how about cordwood? We knew that cordwood masonry was strong on compression and weak on tension; and that 450 gallons of water trying to bust out of a hot tub puts a serious tensile stress on the sidewalls. How could we impart tensile strength to a cordwood tub, so that it would not collapse outward upon filling with water? The answer, again, is my other favorite method of wall building: surface bonding. The technique is described below.

The dimensions of the tub take advantage of a standard blue polyvinyl liner size: 60" in diameter by 48" high, eliminating the expense of having a custom liner made to order. A 6'-diameter liner is also standard, but takes 44 percent more water to fill. And four friendly people can get into the 5'-diameter tub without difficulty, so the larger size didn't seem worth the extra water usage and heating requirements. I strongly advise a cylindrical tub if it is to be made of cordwood, so that stresses will be equal all the way around.

Cordwood construction took place as described elsewhere in this book. We used 8"-thick walls, a convenient thickness for the rather small diameter, and easily obtained by cutting extra 16" log ends exactly in half. Inner and outer mortar joints were each about 2½" wide, with a 2½" sawdust-insulated cavity between. Thus, the walls of our hot tub have an insulation value of about R-8. We used the outlet drain, already laid under the concrete floor, as the center point of the tub, and, with a red crayon, described the inner circumference of the tub using a 30" radius. We were careful to keep any irregularities in the cordwood to the exterior, and kept the inner surface relatively smooth and round by frequent use of a 4' level. The masonry is built up to a height of 42".

It is important to remember to leave spaces for your cold-water return pipe (about 9" off the tub bottom) and your hot water inlet (12" beneath the rim of the tub, on the opposite side of the cold water return). (*See* 23-1) We accomplished this by putting in 8" long by 4" diameter ABS pipes at the appropriate places during construction. Have the pipes cut, and place them near the work as reminders. If you forget to put them in while laying the cordwood, it will be extremely difficult to install them later.

Next, we stapled 1"-mesh chicken wire to the inner surface of the cylinder, not so much for its tensile strength, but more to provide a good receiving surface for the surface-bonding cement. As the

---

* Although the tub has been removed for reasons already cited in Chapter 8, I am leaving the story of its installation and use intact, as it has generated a lot of interest over the years.

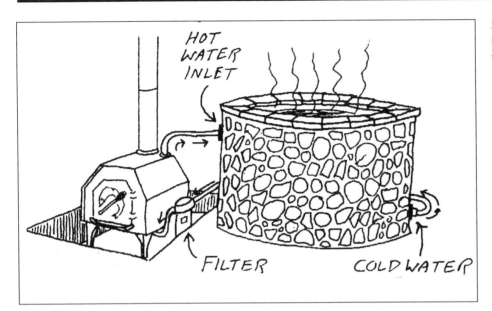

23-1. The cordwood hot tub is heated with a special wood stove.

bonding cement is quite expensive, we used a preliminary parge of the same mortar mix we used with the cordwood masonry, as seen in 23-2.

23-2. The author applies mortar to the chicken wire stapled to the interior of the cordwood hot tub. Next, the surface-bonding cement will be applied to increase the tensile strength of the tub.

With most of the irregularities on the inner surface filled in this fashion, we proceeded to apply a ⅛" coating of surface-bonding cement, giving a very smooth surface with excellent tensile strength. We capped the wall with a strong inch-thick cap of 5 parts sand, 2 parts Portland cement. (*See* 23-3.) Later, the blue polyvinyl liner can be glued to this cap.

The tub is heated by a special wood stove designed for the purpose, manufactured by F. X. Drolet, P. O. Box 178, Limoilou, Quebec, Canada G1L 4V4.

The stove, called Aqua-Flames, has a water jacket on all sides except the front-door panel. It comes with a special cowling and stovepipe for outdoor use, although we installed it indoors. The water is heated in the manifold which wraps the stove and is delivered to the tub by the thermal-siphon principle already discussed.

In order to increase the power of the thermal siphon, we set the Aqua-Flames stove into a rectangular depression 16" into the floor, as seen in 23-4. The greater the height differential between the top of the stove and the hot-water inlet into the tub, the greater the circulation by thermal siphon. The cold-water return takes the cold water off the bottom of the tub and returns it to the stove where it is heated. We have a filter on the cold water return line, just before the water re-enters the stove. It would be possible to add a pump to the line to drive the water faster, but we have found that the water temperature

23-3. Dennis caps the cordwood with a strong Portland cement.

(*See* 23-5 and 23-6.) Our deck units are each made of eight trapezoidal pieces of 2" by 10" pine, glued to a semicircular base with epoxy. Figure 23-7 shows how the two semicircles can be cut from a single sheet of ½" plywood.

After gluing the trapezoids to the semicircles, every surface and edge of the deck pieces is covered with two layers of waterproof epoxy, the kind of clear hard plastic coating made by mixing an epoxy resin with a hardener. When dry, each deck piece is fastened in three places to the tub's cement cap by using a 3" × ⁵⁄₁₆" lag screw in combination with the correct ½" × 1" leaded expansion socket set into the cement. These sockets are also known as *lag expansion shields*. The two deck pieces finish the tub attractively, and they are comfortable for sitting, as well as for entering and exiting the water. They are easily removed for cleaning as needed.

The tub is a joy. Seven-year-old Rohan especially enjoys swimming and playing in it, and Jaki and I

will rise quite rapidly with no pump connected. In fact, the use of a pump causes the firebox temperature to be lower, causing inefficiency and severe creosote creation in the woodburning process.

The inlet and outlet fittings are installed in the blue polyvinyl hot-tub liner at the exact positions to feed through the two 4" pipes built into the cordwood walls. This is tricky work and it may be worth your while to hire the liner supplier to make these connections for you. The hot-and cold-water lines to the stove can be installed using standard 2" plumbing fixtures and pipe.

Our only problem was not getting an absolutely watertight connection at the cold-water return outlet. Once, we had to remove all the water from the tub and fix the leak, but, again there was a tiny leak at the same fitting. Thankfully, in a few days the leak had sealed itself.

Once all the leaks are repaired the top of the liner can be glued or taped to the Portland cement cap—we used Bituthene®, handy for so many different things—and a pair of wooden semicircular laminated deck pieces can be screwed to the cement cap.

23-4. The Aqua-Flames wood stove is set 16" into the floor to provide a greater thermal siphon. Later, we moved the filter to the cold-water return line, as shown in 23-1.

23-5. (left). In addition to epoxy glue, each 2" × 10" trapezoid is fastened to the ½" plywood with a single wood screw, as seen here. Two large staples are used on the top side to help fasten each trapezoid to its neighbor, reducing flexure of the whole unit.

23-6 (right). The 2" extruded polystyrene cover is made in two pieces for easy handling and floats on the water. The insulation value is about R-10.

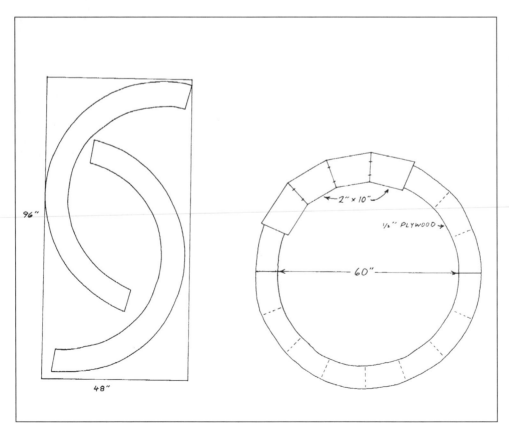

23-7. Two semicircles can be cut from a single sheet af plywood, as shown here. Place the plywood on the tub and mark the semicircles with a pencil from below. Cut the pieces out with a jigsaw. Eight trapezoidal pieces of 2" × 10" are fastened to the plywood, as shown here and also in 23-5 and 23-6.

will often watch the last half hour of a TV movie while soaking. Performance? Excellent. In one test, we heated the water from 66° F to 102° F in 4½ hours. With the 2" extruded polystyrene cover in place (23-6), only about 3° F is lost overnight, so we sometimes use the tub for two days on a firing. The temperature can be brought from 95° F to 102° F in an hour with one good firing of the stove. During firing, an occasional stirring of the water with a canoe paddle serves to destratify the water. (*See* 23-8.)

23-8. Stirring with a canoe paddle destratifies the water.

24

Outbuildings

THE SAUNA

The main cylindrical structure of the sauna was built in July, 1982 by cordwood workshop students from all over the United States and Canada. One student, a surgeon from "Moberly, MO" and a jolly sort of fellow, commented during the workshop: "You know, Rob, I never believed the story about Tom Sawyer and the fence whitewashing until this. Here we are paying you to build your sauna!" At the end of the week, he—and the others—gave Jaki and me high marks on the workshop critique sheet. We finished the roof in the autumn.

The sauna has an inside diameter of 9'. Cordwood work proceeded very much as with the hot tub. We used 8" log-ends of old split rails and sawdust as insulation. Differences in log-end length were averaged between the inside and outside, except at the lower part of the northern hemisphere, which was to be earth-bermed. Here, we endeavored to keep the outside of the wall as smooth as possible for the application of the parging coat of cement plaster. We confirmed our test (see 12-2) that cordwood masonry makes an excellent receiving surface for a plaster parge. Later, we waterproofed this detail with the Bituthene® membrane. In place of the buttresses which we used to resist the lateral load on the main building, we had anchored two 8" corner blocks right to the footing prior to cordwood work. As only about 10 sq ft of the cordwood wall is earth-sheltered, these "mini-buttresses" may not have been entirely necessary. Primarily, earth-berming a small portion of the cordwood sauna was done as a test of the waterproofing technique, and, as such, seems to be a success, at least after nine years.

We applied an inch of Styrofoam® to the wall during backfilling with crushed stone, in order to protect the membrane. A 4" perforated drain tile is an added protection to take hydrostatic pressure off the wall. (See 24-1 and 24-2.)

Outside of several interesting design features incorporated into the wall by some of the more creative students, there was nothing special about construction. At about 5' 8", we set two 8" ceramic thimbles into opposite sides of the building, for use as vents. Log-ends with handles are used to close the gaps when ventilation is not desired. A 3"-diameter piece of ABS plastic pipe is set into the cordwood wall near the floor to supply air for the wood stove. This prevents oxygen starvation for both stove and bathers.

No post is used in the building because of the smaller diameter, but I wanted to stay with the radial rafter design. This presented a problem, as only one rafter could span the building in one piece. The others would have to join to this "primary rafter" in some way. But calculations showed that the primary rafter could not by itself carry its own load as well as the loads of the secondary rafters. One solution, suggested by the good doctor from Moberly, was to greatly beef up the primary rafter. This was carefully considered, and we could have done it by laying two of the four-by-eight rafters side by side. Unfortunately, this solution threw the symmetry of the building off-kilter, and I didn't care for the appearance. I decided to go the complicated route.

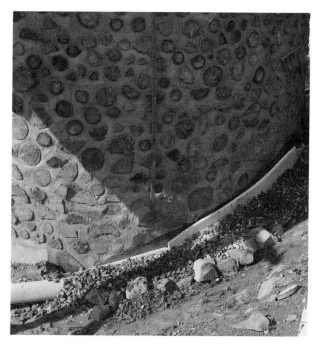

24-1 (left) and 24-2 (right). Styrofoam® is held to the Bituthene® membrane during backfilling with crushed stone. The 4" perforated flexible drain tile comes enclosed with a nylon filtration sock.

THE WAGON WHEEL

After days of thought and discussion, I decided on a wagon-wheel rafter design, using a strong hub system at the center for strength. First, our sixteen 2" × 8" wall plates were set so that the westernmost plate was 2" higher than its counterpart on the opposite side of the circle. Moving around the circle from west to east, each plate is mortared into the cordwood masonry ¼" lower than the previous one. This establishes the roof pitch, which, although slight, is sufficient to promote drainage.

The long primary rafter was loosely positioned, resting on the centerpoints of the eastern and western plates. The six secondary rafters, only half as long, were also positioned loosely, supported by a temporary post. Two octagonal plates, each 16" across, were fabricated of ¼" plate steel. The meeting of the six secondary rafters to the primary rafter was drawn to scale on the metal plates (24-3) and eight ½" holes were drilled through one of the plates. Corresponding holes were then marked and drilled through the rafters, taking care to go through the 8" of wood as straight as possible.

Next, eight pieces of ½" × 10" threaded rod were welded into the holes of the plate, creating a little eight-legged metal table. (*See* 24-4.). The table legs were installed into the holes in the rafters, and the tight fit allowed removal of the temporary post. Now the second plate was marked at the points where the threaded rods came through the bottom sides of the rafters, the holes were carefully drilled, and the bottom plate installed with nuts. (Note: Welding the other end of the rods to the top plate, instead of fastening with nuts, allows the bottom nuts to be tightened as the rafters shrink with the dry sauna heat.) Now the spokes of the wagon wheel are a monolithic unit, joined by the strong double-plate hub system. All of the rafters are now load-supporting. The center of the wheel cannot sag without stretching the bottom plate and compressing the top side of the secondary rafters. Neither of these situations can occur. Two views of the wagon wheel support system can be seen in 24-5 and 24-6.

The snowblock infilling between rafters, also seen in 24-5 and 24-6, was done quickly by using short pieces of 6" milled logs, left over from a friend's log cabin. The logs were laid in the opposite manner of the normal style of cordwood masonry, where the

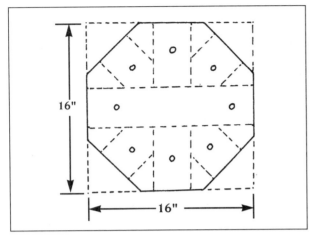

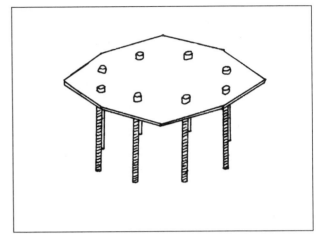

24-3 (left). Octagonal pattern for ¼" thick metal hub plate. 24-4 (right). The octagonal "table." Each leg is a ½" threaded rod welded to the plate.

ends of the logs are exposed. This proved to be a very quick method of filling in the pesky snowblock detail on an 8"-wide wall, and we used a similar method to close off the spaces over the octagonal girder system in the main building. This will greatly decrease sound transfer from the bedrooms and bathroom to the main open-plan living area.

The roof was built up in the same way as the main building, except for one minor difference. Above the planking, we nailed down a layer of ⅜"

chipboard, sealing the joins between sheets with duct tape. There were two reasons for this. Firstly, the planking we used was old recycled silo boards, and the chipboard provided a better surface for receiving the Bituthene®. Secondly, I was concerned that the 200° F temperatures common near the roof of the sauna might cause the Bituthene® to melt and find its way down through the boards. We'd done a similar installation at the Log End sauna without problems, and, after nearly ten years' use, have

24-5 (left) and 24-6 (right). Two views of the wagon-wheel support system.

24-7. The author nails old two-by-six silo boards to the wagon wheel.

experienced no problems at the Earthwood sauna either. Illustrations 24-7 and 24-8 show the sauna roof construction.

A Metalbestos chimney was installed using the appropriate flashing cone for shallow-pitched roofs. Two inches of scrap Styrofoam® and beadboard were laid over the membrane as protection, and then the other layers were built up as at the main building: 6-mil black polythene, 2" of crushed stone, hay, and 8" of earth. The sauna roof, easily accessed and viewed from the deck, can be seen in 20-3.

The sauna benches are made of two-by-six white cedar, planed on four sides. Poplar (quaking aspen) also makes a good sauna bench, as we learned at Log End. Avoid hardwoods, with their high conductivity of heat, and pitchy softwoods.

The door and thermopane windows were installed as already described for the main building.

We heat the sauna with a homemade box stove of ¼" plate steel. I do not recommend a cast-iron stove, because of the thermal shock which would occur when the stove was fired in sub-zero temperatures. We covered the top of the stove with bricks to decrease the direct radiation from the metal to the skin. If steam is desired in the sauna, a little water

can be poured on the bricks... carefully. Temperatures of 180°F and more are quick to obtain and easy to maintain, even in the winter, and the heat is comfortable because the radiation is greatly reduced by the bricks and the atmosphere is dry (if steam is not used).

THE OFFICE

My 20'-diameter office, where I am now writing this book, is geometrically between the sauna and the house in size. The footings and floor were poured monolithically, on a pad of compacted sand, with a single pillar footing located at the exact center of the slab.

Again, the bulk of the wood masonry took place during our Cordwood Workweek in July of 1983. The versatile George Barber supplied us with a 29'-diameter hemispherical geodesic dome of his own design, and, along with his daughter and son-in-law, helped us to erect it over the building site. The entire shed, including the roof with its 12" over-

24-8. Low-cost chipboard covers the rough planking, the joins sealed with duct tape. Flashing is installed around the perimeter. The Bituthene® membrane covers the roof.

24-9. George Barber strikes again! The entire shed can be built under the cover of this 29'-diameter hemispherical dome.

THE COLUMN AND CAPITAL

For the central column, we used a 12"-diameter hemlock post obtained by barter from a neighbor. A 9'-tall post of this type will support 60 tons, ten times our load. We positioned the base of the post on the concrete floor by employing two lag screws which threaded into lead receiving sockets set in the floor. This technique has already been described in detail in Chapter 5.

At the Parthenon in Athens, the ancient Doric columns are topped with elaborately carved *capitals*, which, among other things, serve the function of increasing the bearing surface of the top of the column. The capital at the Earthwood shed is very much more rustic: a 24" diameter octagon made up of pieces of a six-by-twelve beam, as shown in 24-10 and 24-11. Each of the sixteen five-by-ten rafters, then, has plenty of bearing surface. The outer end of the rafters is supported by a system of sixteen pairs of 2" × 6" plates.

Illustration 24-12 shows the capital in position. The capital is fastened to the top of the post with two

hang, can be constructed under the cover of this huge umbrella, after which the dome is dismantled for use again somewhere else (24-9).

Structurally, the shed differs slightly from both the sauna and the Earthwood house. The sauna has no internal post, thanks to the wagon-wheel roof system. The house has the stone mass at the center for support, as well as the octagonal post-and-girder system to shorten the rafter spans. The wagon-wheel idea was out of the question at the shed because of the 18' clear span, yet the building is too small to accommodate an ancillary support system like the octagon. The obvious answer was a single sturdy post in the center, but this raised another problem: How could a post any smaller than 2' in diameter provide enough bearing space for the joining of the 16 heavy five-by-ten rafters required to support the earth roof? We had to borrow from the ancient Greeks to solve this one.

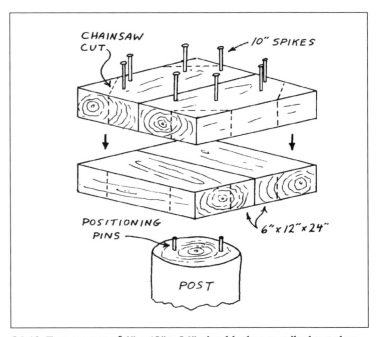

24-10. Two courses of 6" x 12" x 24" pine blocks are spiked together to form a large block one foot thick and two feet square. The octagonal capital shown in 24-11 is cut from this large block with a chain saw. The capital is pinned over a 12"-diameter post.

24-11. The sixteen five-by-ten shed rafters meet over the capital, made as shown in 24-10.

positioning pins. We used log-cabin spikes with the heads cut off. The use of two positioning pins at each join, instead of one, prevents the post and the octagonal capital from rotating during installation of the rafters.

I have taken the time to describe this post-and-rafter configuration in some detail, as I feel that the shed building would make an excellent starter project for the inexperienced owner-builder. The building, with its 254 sq ft of usable area, would serve as an excellent temporary shelter to live in during the months of the main house construction. Besides the elimination of interim shelter costs, the strategy of building a small temporary shelter yields valuable building experience and a useful outbuilding for the future.

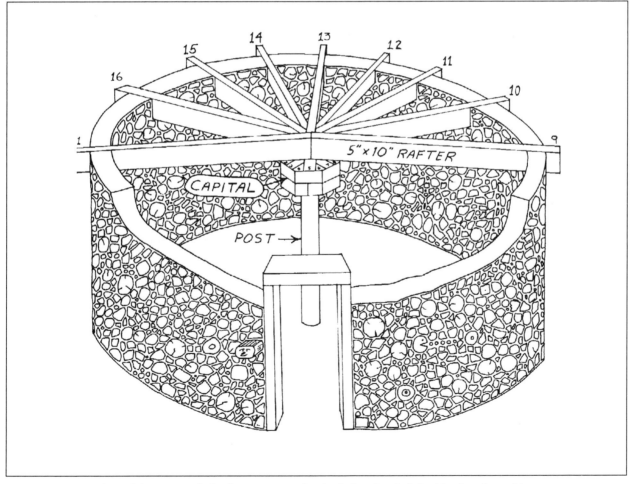

24-12. A cutaway view of the post-capital-rafter system at the shed, showing 9 of the 16 rafters in position.

24-13. The office at Earthwood is 20' in diameter.

Of course, if you happen to have a 24"-diameter tree trunk lying around, you can eliminate the column and capital detail described above. Your only problems will be hauling the thing to the site and tipping it up into position. (We actually did this at Mushwood.)

The dome which we used as a temporary cover while building the office eventually became our garage, tucked away in the woods along the top driveway. The finished office is shown in 24-13.

THE SHED (ROHAN'S BUILDING)

George loaned us another dome to cover a new worksite. This building was to be Rohan's hangout, but, so far, he hasn't made much use of it that way, and it has become a storage shed, the original intent for the office building.

The shed is 16" outside diameter with 12" round cedar log-ends. It is built on—what else?—a monolithic floating slab. The radial roof system is composed of sixteen four-by-eight rafters. Instead of planking, Rohan and I installed a ¾" plywood deck, which has no trouble supporting the earth roof. The mortar is the Portland mix described in Chapter 2. The building cost about $1000 in materials. And that cost for the building even includes the wood stove and Metalbestos chimney, windows, and a door made of two-by-six planking. This works out to about $6.50 per sq ft. Like my office, it is easy to heat with the wood stove.

25

Garden, Landscaping, Stone Circles

Stone circles? Yes, wait for it.

Reclamation of waste land, such as the gravel pit in which we built, is either an expensive project or a long-term labor-intensive project. We have adopted the second course. Still, in ten years we have done quite a bit. And nature has done some reclaiming of her own.

THE GARDEN

Jaki gardens by the biodynamic French intensive method, by which a lot of vegetables are grown in a small space, especially in *raised beds*. Our four beds at Earthwood are each 4' wide by 10' long, a total of just 160 sq ft of garden.

The beds are naturally rot-resistant cedar logs milled on three sides, and lined with aluminum flashing, so that the wood surrounds will last even longer. Also, the bottom courses rest on crushed stone, not soil. The beds are over a foot deep and we keep them well supplied with nutrients such as rotted manure and compost turned in each year. We like these "permanent" beds, and feel that there is less work in the long run, but we've seen excellent gardens of mounds reshaped at the time of turning in the spring. The advantages of raised beds are:

25-1. Jaki's garden, just after harvest, 1991.

229

(1) Less land is needed. Intensive planting is possible as no space between rows is required, thereby yielding much more produce per square yard of garden.

(2) Seeds or seedlings are spaced so that the young plants form a living "green mulch" over the bed, discouraging weeds and helping to retard moisture loss through evaporation.

(3) Gardening is three-dimensional: Leafy vegetables are alternated with root vegetables so that both the surface and the subsurface are productive.

(4) Each bed can be tuned to the proper pH factor (acidity vs. alkalinity) for the particular vegetables to be grown in that bed. Lettuce likes sweet soil, for example, while strawberries prefer acidic soil. Certain special mineral requirements also can be economically satisfied by concentrating them where needed.

(5) Much less watering is required, because of the lack of runoff and the reduction of evaporation waste. This advantage is greater with permanently bordered raised beds than with the mounded method.

(6) Much more economic use is made of mulch and compost with intensive gardening. Good soil is built up faster.

(7) No rotary tiller or cultivator is required. The beds are easily maintained with a spade and small hand tools. The raised beds are easier to work on because they are 8" to 10" above the permanent walkways.

(8) Bug control is easier.

(9) Raised beds are aesthetically superior. They are neater and tidier, due to the permanent walkways, and very little weeding is required.

(10) In combination with a built-on protective cover and the use of rigid foam insulation around the inside edge of the bed, growing seasons can be greatly extended. Leandre and Gretchen Poisson, who pioneered this technique, grow certain vegetables year-round this way in snowy New Hampshire.

Their Solar Pod® raised bed cover (25-2) can extend the gardening season by several weeks. The Poissons have also pioneered a do-it-yourself masonry stove system and several true passive solar housing innovations, such as their Sun Eye windows. Write: Solar Survival, Box 250, Harrisville, NH 03450.

Our garden is easily watered with the bicycle pump system. We connect a 50' hose to a hose bib near the solar room, pedal to build up a 25-gallon reserve under pressure, and away we go. Or one of us waters while the other pumps. Someday, we hope to make use of rainwater by way of barrels under the downspouts and a wicking system. And we wouldn't mind doubling the number of raised beds, although Jaki grows quite a bit in just 160 sq ft.

The tennis court and pond we once imagined adding are still dreams. We drag the tennis court idea out of the closet every once in a while. And there is still a place for it, a yet unreclaimed flat section of gravel pit about the size of ... a tennis court. In the meantime, we have planted a lot of grass, with a good flat area in front of the house for volleyball and similar activities. But one landscaping activity unforeseen ten years ago was the...

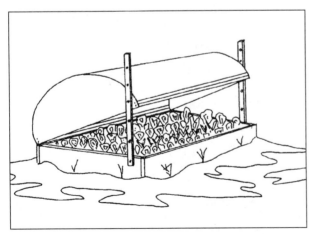

25-2. The Solar Pod®, developed by Leandre and Gretchen Poisson.

STONE CIRCLES

I've long had a fascination with megalithic building technology, and have visited all the great stone circles in the British Isles—Stonehenge, Callanish, Brodgar, Avebury—some of them several times. There were some beautiful stones left behind in the gravel pit many years ago, and I found out that George Barber's daughter owned the remains of an old granite quarry just a mile down the hill. An idea began to grow in the back of my head, but it took a great leap forward when I actually visited the old quarry and found several beautiful long anorthosite stones just lying there, abandoned 150 years ago like the Easter Island carved

25-3. The stone circle is laid out in Megalithic Yards, like the ancient circles in Britain. To give a sense of scale, the outside diameter is about 30'. The Bunny Stone is at ten o'clock in the picture.

heads still lying in their quarry. Marie Barber graciously told me to help myself. Eight massive stones from Marie's quarry formed the nucleus of our stone circle, built in 1987. One of them looks like a rabbit sitting in the grass. We call it the Bunny Stone.

There are actually two stone circles, the main set of concentric rings which appear in 25-3, and another minor circle, called the "Children's Circle" about 100 feet away. Peter Allen and I, aided by his small backhoe, built the inner ring of sitting stones and the Children's Circle in about a day. The twelve standing stones, however, would require a bigger machine and a professional operator. Who better but old Ed Garrow from way back in Chapter 9? In two days, Ed removed the eight stones from the quarry (and several other beauties from around the hill), hauled them to the site, and stood them up in sockets in predetermined locations.

The largest stone, weighing almost two tons, we deliberately left for last when Ed shut the backhoe down for the day. Eight of us, led by my old friend Russ Keenan, moved the stone by hand, about 40 feet, using only rollers and levers; we imagine just

25-4. The stone circle completes the kraal at Earthwood. Rohan, on left, works to complete the landscaping.

like Stone Age man. Then we almost stood it up. Almost. But it got too dark and there was some danger involved, so we left it for the night. The next morning, our crew was down to two—Russ and me—and the backhoe was still sitting there. We decided that discretion was the better part of valor, and we had the operator set the stone before departing with the machine. We came away with a whole lot of respect for the megalithic builders, who stood up stones perhaps 40 times as big as ours, without benefit of heavy machinery.

The circle has delighted hundreds of visitors at bonfires over the years. It is all laid out in precise celestial orientations and the major compass points, thanks again to George Barber. And four outlier stones, quite a distance from the main circle, mark the rising sun on the longest day, and the setting sun on the shortest day. At $500 for heavy equipment work—the stones were free—we have definitely gotten our money's worth. The circle seems to balance the other round buildings, and complete the "kraal" at Earthwood, architecturally and spiritually (25-4.).

Earthwood Photo Album

A. The original three buildings at Earthwood (office, house, sauna) as seen from the stone circle. Rohan's building is out of view to the left.

B. "Welcome to Earthwood"

C. The earth berm extends 13' up the northern hemisphere at Earthwood.

D. The office and main building at Earthwood.

E. Large window wells increase the size of the conversation area.

F. Jaki likes her kitchen.

G. The wood chute leads to the firewood storage area downstairs.

H. The bedroom has a medieval feeling.

I. Jaki and Rohan enjoy a dip in the cordwood hot tub.

J. The round sauna at Earthwood.

APPENDIX ONE

On Big Projects

For several years I've known the dangers of tackling big building projects, having observed several marriages and personal relationships go sour during the seemingly never-ending construction process. Living in a tent or cramped shed for too long, or, worse, trying to live in a half-completed house itself, is just too stressful for many people to handle... maybe most people. I'd even written a lengthy discourse on the wisdom of building small in my previous book. But, with the experience of two houses behind me, I thought I could ignore my own advice and lunge forth into the massive Earthwood project with impunity. At workshops, I sometimes wear a T-shirt adorned with a sophomoric old guru sitting in the lotus position. The caption is: "Take my advice... I'm not using it. " It wouldn't be so funny if it weren't so true.

Jaki knew better. She was not surprised when the pressures of housebuilding began to take their toll on my equanimity. But neither of us was prepared for the discouragement and depression that came with the hardwood-expansion debacle. And the old familiar scenarios returned: rushing to beat the winter, financial pressures, and just plain disenchantment with the lengthy building process. Two things got us through the project with our marriage intact: Jaki's patience, and the fact that we had a comfortable home to return to each evening.

We moved into Earthwood thirteen months after breaking ground. The upstairs, a complete apartment in and of itself, was comfortably finished, although we had to draw water from the outdoor pump for five months prior to installing the bicycle-powered water system. Completion of the downstairs took another eighteen months. When the pressure is off, work slows down dramatically.

I feel strongly that a project on the scale of Earthwood is too much for inexperienced owner-builders to tackle, not because the techniques are too difficult; they aren't. The house is simply too big. A one-storey version of the building, however, would be well within range. It would even be possible to build the lower storey one year, cap it and live in it for a while, and add the second storey another year. For many, this may work out to be the affordable route, too, especially if the goal is to avoid bank financing. Another good plan is to build the shed first, for practice and to learn how long it takes you to do things.

Conclusions

FUNCTION

Despite the somber tone of Appendix One, our experiment in integrative design has been a success. Earthwood is more energy-efficient than we'd anticipated, using only 3¼ full cords of dry hardwood per year for heat. There is plenty of room for work, recreation, storage, food production and preservation, and home industry.

COMFORT

Earthwood is fun to live in, and comfortable both physically and spiritually. I credit some of this to the advantages and sensations of living in the round, but there is also a good feeling present in almost any home which makes use of strong natural materials such as logs, stone, heavy timbers, and earth.

COST

Inclusive of hired labor, the *total cost* for Earthwood, by the time of our moving in (1982) came in at about $24,000—or about $12 per sq ft. About two-thirds was materials, one-third was hired help. Although some of the materials were *donated*, their value is included in the $24,000 figure. *Not included* is the land or the energy systems. Since 1982, I'm sure we've spent another $3000 on such things as floor covering, new interior walls because of use changes, and improved plumbing and electric work.

LABOR

Together, Jaki and I have invested between 2500 and 3000 hours of our lives in the project, including time spent building walls and tearing them down again. We paid for the building as we progressed, so there were periods when little work was done for lack of funds, or we would do labor-intensive jobs such as bleaching beams, sanding the floor, or painting interior walls.

WAS IT ALL WORTH IT?

There were times when we did not truly know the answer to this question, but we have no doubt now. Earthwood is really the culmination of years of study and work in the "alternatives" field, and we are well pleased with the results. It is entirely possible that we will spend the rest of our lives here. On the other hand, one never knows what the future will bring. It may be that we have many paths to tread in this life, and miles to go before we sleep.

Bibliography

CORDWOOD MASONRY

*Flatau, Richard. *Cordwood Construction: A Log-End View.* Merrill, Wis.: Self-published, 1983.

*Henstridge, Jack. *Building the Cordwood Home.* Upper Gagetown, N. B., Canada: Indigenous Materials Housing Institute, 1978.

Northern Housing Committee. "Stackwall: How to Build It." Winnipeg, MN, Canada: University of Manitoba, 1977.

Roy, Robert L. *Cordwood Masonry Houses: A Practical Guide for the Owner-Builder.* New York: Sterling Publishing Co., 1980.

———*Earthwood: Building Low-Cost Alternative Houses.* New York: Sterling Publishing Co., 1984.

EARTH SHELTERING

*Campbell, Stu. *The Underground House Book.* Pownal, Vermont: Garden Way Publishing, 1980.

*Oehler, Mike. *The $50 and Up Underground House Book.* New York: Mole Publishing Co. and Van Nostrand Reinhold Co., 1978.

Roy, Robert L. *Underground Houses: How to Build a Low-Cost Home.* New York: Sterling Publishing Co., 1979.

*Underground Space Center, University of Minnesota. *Earth Sheltered Housing Design.* 2nd ed. New York: Van Nostrand Reinhold Co., 1985.

GENERAL

Cole, John N., and Charles Wing. *From the Ground Up.* Boston: Little, Brown & Co., 1976.

———*Breaking New Ground.* New York: Atlantic Monthly Press, 1986.

Gordon, J. E. *Structures, or Why Things Don't Fall Down.* New York: Plenum Publishing, 1978.

*McClintock, Mike. *Alternative Housebuilding.* New York: Sterling Publishing Co., 1984.

Roy, Robert L. *Money-Saving Strategies for the Owner-Builder.* New York: Sterling Publishing Co., 1981.

*Sobon, Jack and Roger Schroeder. *Timber Frame Construction: All About Post-and-Beam Building.* Pownal, Vermont: Garden Way Publishing, 1984.

*Sterling Publishing Co. *The Encyclopedia of Wood.* New York: Sterling Publishing Co., 1989.

*Thoreau, Henry David. *Walden.* New York: New American Library, 1960.

*Books marked with an asterisk (*) are available through Earthwood Building School, RR 1, Box 105, West Chazy, NY 12992.

Note: My previous books (Roy, Robert L.) are out of print, but may be found at libraries.

Source Notes

PART ONE

1. Square, David, "Poor Man's Architecture," *Harrowsmith* No. 15 1978, pp. 84–91.
2. Jenkins, Paul B., "A Stovewood House," *Wisconsin Magazine of History*, December, 1923, pp. 189–193.
3. Tishler, William H., "Stovewood Construction in the Upper Midwest and Canada: A Regional Vernacular Tradition," a paper appearing in *Perspectives in Vernacular Architecture*, Annapolis, Maryland: Vernacular Architecture Forum, 1982, pp. 125–136.
4. Perrin, Richard W.E., "Wisconsin 'Stovewood' Walls: Ingenious Forms of Early Log Construction," *Wisconsin Magazine of History*, Spring, 1963, p. 216.
5. Ibid.
6. Tishler, op. cit., p.133.
7. Henstridge, Jack, *Building the Cordwood Home*, Indigenous Materials Housing Institute, Upper Gagetown, N.B., Canada, p. 5.
8. Flatau, Richard C., *Cordwood Construction: A Log End View*, self-published, Merrill, Wis., p. 83.
9. Henstridge, op. cit., "Update" to 6th printing.
10. Henstridge, op. cit., p. 41.
11. Square, op. cit., p. 87.
12. Flatau, op. cit., p. 122.
13. Henstridge, op. cit., p. 35.
14. Hodges, Tom, "The Peeling Spud: A Handy Tool for the Homestead," *The Mother Earth News*, July, 1976, p. 122.
15. Cole, John N. and Charles Wing, *Breaking New Ground*, The Atlantic Monthly Press, New York, 1986, p. 139.

PART TWO

1. Thoreau, Henry David, *Walden*, New American Library, 1960, p. 13.
2. Ibid., p. 15.
3. Ibid., p. 215
4. Wells, Malcolm, "Underground Architecture," *The Co-Evolution Quarterly*, Fall, 1976, p. 87.
5. Burns, Robert, poem: "To a Mouse."
6. Gordon, J.E., *Structures: Why Things Don't Fall Down*, Plenum Publishing, 1978, p. 172.
7. Cole, John N. and Charles Wing, *From the Ground Up*, Little, Brown & Co., Boston, 1976, p. 129
8. Haynes, B. Carl, Jr., and J.W. Simons, *Construction With Surface Bonding*, Agricultural Research Bulletin No. 374, U.S. Dept. of Agriculture Research Service.
9. Ramsey, Charles G. and Harold R. Sleeper, *Architectural Graphic Standards*, 7th Printing, John Wiley & Sons, 1963, p. 29., Elwyn E. Seelye, consulting engineer.
10. Barden, Albie, "Masonry Heaters: On the Comeback Trail," *Alternative Sources of Energy*, Sept.–Oct., 1983, p. 19.
11. Cole and Wing, op. cit., p. 124

Index

KEEP IN TOUCH, SHARE THE WEALTH

This certificate is awarded to students at Earthwood Building School and other cordwood builders who send us a picture of their project, preferably with a short commentary with any details of construction to contribute to the database. The certificate entitles the holder to put the initials M.M.S. (Master Mortar Stuffer) after his or her name, which then makes that person just as fancy as anyone else. And, unlike many other lesser degrees and honors awarded, it shows that the recipient actually went out and did something to earn it, something good for the individual, his or her family, and the planet. Of course, the structure already proves that, but everyone likes a little recognition.

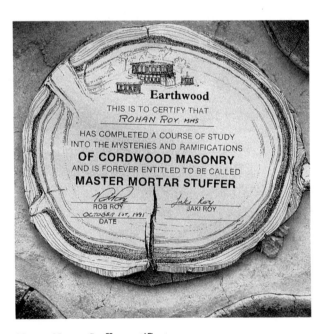

Master Mortar Stuffer certificate.

Rob Roy.

About the Author

Rob Roy was born in Webster, Massachusetts in 1947. He left the American middle-class life-style in 1966, traveling to over forty different countries in two years while living out of a backpack. He finally settled in Scotland where he met Jacqueline Bates, now his wife. In 1974, the Roys moved to West Chazy, New York, and built Log End, their first homestead. In 1980, they refined their building techniques a half mile away at Earthwood and started the Earthwood Building School, specializing in cordwood masonry and earth-sheltered housing.

This is Rob's sixth book in the alternative building field. He has written articles for *The Mother Earth News, Alternative Sources of Energy, Farmstead, Fine Homebuilding*, and others. With their sons Rohan and Darin, Rob and Jaki pursue a self-reliant life-style on their homestead, making electricity from sun, wind, and pedal power, heating with wood, and growing vegetables in their garden.

Metric Conversion

INCHES TO MILLIMETRES AND CENTIMETRES

	MM—millimetres			CM—centimetres			
Inches	MM	CM	Inches	CM		Inches	CM
⅛	3	0.3	9	22.9		30	76.2
¼	6	0.6	10	25.4		31	78.7
⅜	1O	1.O	11	27.9		32	81.3
½	13	1.3	12	30.5		33	83.8
⅝	16	1.6	13	33.0		34	86.4
¾	19	1.9	14	35.6		35	88.9
⅞	22	2.2	15	38.1		36	91.4
1	25	2.5	16	40.6		37	94.0
1¼	32	3.2	17	43.2		38	96.5
1½	38	3.8	18	45.7		39	99.1
1¾	44	4.4	19	48.3		40	101.6
2	51	5.1	20	50.8		41	104.1
2½	64	6.4	21	53.3		42	106.7
3	76	7.6	22	55.9		43	109.2
3½	89	8.9	23	58.4		44	111.8
4	102	10.2	24	61.0		45	114.3
4½	114	11.4	25	63.5		46	116.8
5	127	12.7	26	66.0		47	119.4
6	152	15.2	27	68.6		48	121.9
7	178	17.8	28	71.1		49	124.5
8	203	20.3	29	73.7		50	127.0